Goshawk Summer

The Diary of an Extraordinary
Season in the Forest

James Aldred

Elliott&Thompson

NOTE

Much of *Goshawk Summer* is based upon field diaries I kept while filming
for *The New Forest: The Crown's Hunting Ground*, a fifty-minute wildlife
documentary commissioned by Smithsonian Channel (USA) and Terra
Mater Factual Studios (Austria). The film was produced by Big Wave
Productions here in the UK and was broadcast on the Smithsonian Channel
in late 2021.

First published 2021 by
Elliott and Thompson Limited
2 John Street
London WC1N 2ES
www.eandtbooks.com

This paperback edition published in 2022

ISBN: 978-1-78396-640-0

A portion of the royalties that the author receives from sales of this book
are being donated to Marie Curie (Charitable Trust Number 207994), which
offers care and support through terminal illness.

9 8 7 6 5 4 3 2 1

A catalogue record for this book is available from
the British Library.

Typesetting: Marie Doherty
Printed by CPI Group (UK) Ltd,
Croydon, CR0 4YY

ks about the natural

nother elusive wood-

Stephen Moss, auth.

'One moves from the close world of filming at a nest to the broader
scene with good effect. But this wildlife account is fascinating in its
own right and well worth reading – it's just that the global pandemic
adds to it … This book opens the world of Goshawks to me in a
way that I enjoyed and will never have the opportunity to experience
myself. And I am pretty sure that will be true for you too.'

Mark Avery, author of *Remarkable Birds*

'Really gets into the heart of the area . . . not just an enjoyable book
to read, but one to learn from.'

Bird Watching

'I was struck throughout by the power and visceral beauty of Aldred's
prose. He writes almost like a poet, placing such emphasis on using
precise and beautiful vocabulary . . . *Goshawk Summer* is such a valuable
addition to the canon of nature writing . . . This expansive and honest
memoir, from a markedly different perspective, is sure to be of interest
to so many readers, and I cannot recommend it enough.'

Kirsty Hewitt, *NB*

'A wonderful insight into natural history filmmaking.'
The Bay

'Enchanting ... the goshawks are the book's central characters but there is much else besides not least the foxes and curlews captured by Aldred's lens.'
Country & Town House

'Highly readable and informative ... In *Goshawk Summer*, Aldred succeeds in bringing his childhood home to life ... This book has opened my eyes to [the New Forest's] scale and richness.'
The Pilgrim

In loving memory of my father, Chris.

Sit Still, Look Long and Hold Yourself Quiet

Arthur Cadman, *Dawn Dusk and Deer*

New Forest, England

Spring 2020

A LOUD CALL SHATTERS THE PEACE. NOT THE BLUNT MEW-ing of a buzzard, but the piercing cry of something infinitely more predatory: a wild goshawk. It echoes through the woods around me. Strident, commanding, forceful. A regal sound for a regal bird.

I can't see her but know she's flying towards me through the trees. She's coming in fast and there's only seconds before she explodes into frame.

I roll camera just in time to catch her landing on the nest. Powerful legs held out in front; a squirrel's limp body clenched in her yellow fist. The chicks clamour for food and a heartbeat later they're rewarded with morsels of flesh plucked from the warm carcass.

The goshawk. Steel grey, the colour of chainmail. Sharp as a sword. A medieval bird for a medieval forest. A timeless scene. The wood holds its breath, the only sound the begging of the chicks and the gentle breeze sieving through trees. The forest hasn't been this peaceful for a thousand years.

I grew up here. Made friends, climbed trees, slept rough on the heath and camped in the woods, but I've never known it like this. There isn't another soul around and while Covid grips the outside world, the New Forest blossoms in a spring like no other. Nature's been given the space to unfurl her wings and they are shimmering.

There are many terrible things to remember about the spring and summer of 2020, but I was one of the lucky ones. With permission to film in the New Forest, lockdown gave me a once-in-a-lifetime opportunity to observe the wildlife of a unique place in a unique moment never to be repeated.

This is a tale of reawakened passions for a familiar childhood landscape now struggling to cope with the pressures of the modern world. A portrait in time, as seen through the eyes of the wild creatures relying on it for their survival.

Above all, it's the story of how one family of goshawks living in a timeless corner of England shone like fire through one of our darkest times and how, for me, they became a symbol of hope for the future.

Monday 6 April

A familiar, welcome sound makes me look up from my phone. The first two swallows of the year have just this moment arrived home from South Africa. Dishevelled and visibly exhausted from their long migration, they perch on the telegraph wire across the road from our front garden. Long sceptre wings half-raised as they diligently preen, running breast feathers through beaks while chatting to each other in a constant bubble of liquid conversation. There's a buoyant urgency to their talk: like a couple of long-haul pilots relieved to have made a safe landing after an arduous flight. They are perched directly above our neighbour's farmhouse, a seventeenth-century collection of tiled roofs and sturdy stone walls moulded from the fertile Somerset earth upon which it stands. I wonder how many generations of swallows their barns and outbuildings have provided refuge for over the centuries. Thousands, I shouldn't wonder. It's almost certain these latest two voyagers were hatched and raised here.

I've not been home long myself. A fortnight ago I was filming the daily fortunes of a family of cheetah in East Africa. Four young cubs and an impressively stoic mother whose job it was to keep her boisterous offspring hidden and safe from the murderous attention of lions, while hunting daily to provide them with meat. Meat that still pulsed as it was devoured in the long grass beneath the bushes. Young impala were a speciality of hers. I'd filmed her using the movement of wind on the grass as cover before accelerating into a blur. Stretching out into a searing streak of intent that ended in a cloud of dust, a throng of thrashing hooves, and a cameraman with a heart rate of 140 beats per minute.

Returning to the kill the following morning, we'd disturbed a warthog and a couple of vultures. In the twelve hours of darkness since we were last there, the carcass had been stripped. Picked clean; an empty cage of bones exposed to the red sky above. The impala's gaping skull was still attached to a spine bent double; neck broken. Sinews clung to bony crevices and a swarm of fat black flies rose to greet us. Then something truly unforgettable happened. Without warning, the air around us was filled with the joyous swooping and chattering of European swallows. Buffeting their way north across the equator, they'd been drawn down to the feast of flies. They wouldn't stay long, but for a few glorious hours they swirled in the wake of our land cruiser as we lurched slowly through the long grass looking for our cheetahs.

I would love to think that these same birds would soon be arriving back in the spring pastures of Somerset, but they were more likely following the Great Rift Valley and Nile up into Eastern Europe. Perhaps making their way to an old barn in the forests of eastern Poland or Belarus.

I continue scrolling through the morning headlines. The prime minister's been in hospital for ten days. The police have no PPE to keep them safe on the streets during lockdown, and reports of domestic abuse are increasing. There's not much good news around at the moment and yet it takes a conscious effort to pull myself away from the bulletins.

Across the other side of the valley, a solitary HGV heads south on the M5, but otherwise the motorway's deserted, as is the normally busy road running through our village. The air is still. A goldfinch serenades the sky from the top of our apple tree and dunnocks squabble in the lilac. The blue sky is clear, not an aeroplane to be seen or heard. All is peaceful and serene.

The spell is broken by the incongruous sound of someone singing 'All You Need is Love' at the top of his voice. He's cycling slowly up through the village towards our house and I can't resist peeking through the hedge. An elderly man in overalls is weaving in and out of the dotted white lines like a kid, clearly enjoying himself in the absence of any other traffic.

The perfect antidote to this morning's depressing news. The swallows look on with avian indifference.

I'd returned from Kenya on 15 March. My last day with the cheetahs had finished with the family silhouetted against a smouldering sunset. The mother sitting upright, scanning the western horizon with her painted eyes while the cubs rolled and pounced on each other in the grass at her feet.

The night flight out of Nairobi had been rammed with foreign nationals trying to get home ahead of the chaos about to hit. A TV in the departure lounge was blaring out its rolling-news update from CNN. It soon became clear the presenters didn't know what the story was. Something bad was on its way, but no one seemed to understand exactly what. I certainly didn't. I'd had my head in the sand and was now struggling to re-engage with a world that seemed just as confused, panic-stricken and helpless as the young gazelle I'd seen gripped by the throat the day before. Boarding the plane, I passed a sobering and divisive poster tacked to the wall: 'The New China Virus. What do we know about it?' Dangerously little, it seemed.

I glance up. The swallows have gone, but later I see them swooping low over the meadow next to us, indulging in some serious in-flight refuelling. They've only been back a few hours, but already they look better. Round here they're known as bluebirds, and the sun shimmers on their cobalt wings as they do what they do best, raising the spirits of anyone taking the time to watch. The bluebirds are back, and everything is going to be all right – I hope.

Friday 10 April

The country's been in lockdown for two weeks. I take our three boys into the empty landscape of the valley opposite for some decompression. They've been bouncing off the walls at home and it's good to feel the stride of open ground. They bring their bows to shoot arrows high into the sky above the wide rhyne-locked levels. It's a good way to let off steam for an hour or so. Crossing one of the many small bridges, I glance down to see the five-toed pads of a large dog otter imprinted in the soft mud. A quick peer over the other side of the bridge shows an inky black smear of spraint on stone. I dab in the tip of my hazel walking stick then offer it up for the lads to smell: pungent ammonia with a tinge of weed and fish. They wrinkle their faces and ask why I'm so excited. 'It's just poo, Dad.' I struggle to give them a satisfactory explanation. Homeschooling at its best.

It becomes clear that it's not just us decompressing out here. Nature is also filling her lungs, expanding into the newly reclaimed space of an empty English countryside. We disturb several roe deer and catch the whirling red propeller of a fox's tail as it sprints into a thicket of willow. We encounter a stoat bounding down the track towards us, then stop for a while to tune in to the electric buzz and interference of a sedge warbler. Just like the swallows, these tiny birds have spent the last few weeks flitting their way back north over the sands of the Sahara.

The reed beds are now alive with their song as they seek to establish breeding territories. Me: 'They spend the winter in Africa, lads.' Tarun, our nine-year-old, muttering as he walks past: 'Bet they wish they'd stayed there.'

The boys have just discovered a large clutch of ten pheasant eggs in the leaf litter below a holly bush when my phone rings. It's Andy Page, head keeper of the New Forest. Corralling the lads away so that the perfectly camouflaged hen can get back to business, I answer the call.

'James, bad news I'm afraid. That gos nest isn't sitting. I've just been over to check and there's no sign of her.'

As one of the forest's top natural predators, goshawks have been chosen as major characters for our film. Being rare, legally protected and notoriously elusive, these 'phantoms of the forest' can be a real challenge. Much to my relief, a month or so ago Andy had offered to show me an old nesting site as a potential location for filming. There's not a lot he doesn't know about these legendary raptors and the site he had in mind was perfect. A discreet territory, shadow-locked deep inside a large block of mixed conifer. So, two days after I'd arrived back from Africa and a week before lockdown began, I'd driven south from Somerset to meet him at his cottage on the edge of the New Forest. We sat in his dining room, discussing plans over a cuppa while an ancient stuffed curlew eyeballed us from within a glass case in the corner of the room.

Andy carries his formidable knowledge lightly. A devout birder, he is also the Head of Wildlife Management for Forestry England South, which means it's his job to know what is breeding where and when in the forest. It's also his job to manage the forest in a way that balances the needs of its visitors and residents with those of the wildlife that calls it home.

Siskins were flitting around the garden feeders as we left his cottage to drive out into the forest, yellow dust filling the air as we crossed the stream where as a teenager I walked our dog. This is a landscape of intense memory for me. I left the forest many years ago, our family home dissolving in the wake of my parents' divorce. My sister started a new life in Australia, while I headed to Bristol. To this day, part of my heart remains in the forest, dwelling in the quiet rides and woods of my childhood. Even the smell of the place stirs deep currents of longing within me.

Andy drove with the confidence of someone who knows the exact location of every axle-breaking pothole. Vehicle access to the forest is restricted via gates that function like a series of airlocks leading deeper and deeper into the woods. Having stopped his truck to unlock the first, he led me through the trees to a second, before recrossing the stream next to the old oak beneath which my wife and I spent a lazy summer's day as teenagers.

Entering the next enclosure, we moved up through a large block of forest to emerge onto open heath, where a dishevelled man in hoodie and jeans was stumbling through the young pines next to a secluded car park. Apparently searching the ground for something, he visibly blanched at the sight of an official 4×4 and as Andy walked past my open window to relock the gate, he whispered, 'There's a lot of strange behaviour goes on in these car parks, James.'

Fifteen minutes later, I was standing on the edge of a quiet forest track in the heart of a remote stand of conifers. Goldcrests called unseen from above as I quietly followed Andy uphill through tall, straight trunks of mature Douglas firs. The occasional larch was among them, but their delicate needles were not yet showing, so they looked almost dead in comparison with the evergreens above. There was an eerie silence in that wood. The mossy ground soaked up light and sound like a sponge and no birds were singing. The air was cool, chilly even, and above our heads the wind breathed through the foliage in hushed whispers. The place felt empty, deserted, yet pensive. As if something was watching.

Picking my way forward between fallen branches, I began to notice wood-pigeon feathers and splashes of chalky-white raptor dung. The mutes seemed fresh, but other bird sign was sparse, no more than a thin veneer of ephemeral hints and clues. I wondered how Andy could be so certain we were standing in the heart of a goshawk territory.

In answer to my thoughts, the brooding silence was broken by a strident *kek-kek-kek* that dispelled all doubt. Being mid-March, it was still too early in the season for birds to be on eggs, but it was a clear message: a pair of goshawks had staked a claim to the wood; they had seen us and they didn't much care for the intrusion. As the calls trailed off into the distance, we continued on, up through the trees.

Nearing the top of the slope, Andy paused, raising his binoculars to look at a large bundle of sticks high in a bare, skeletal larch.

It always amazes me how prominent goshawk nests are when you finally find them, but then goshawks are the ultimate paradox: secretive yet bold; skulking yet brazen. Shrouded in shadow, they have an inner fire that burns white hot. An uncompromising, relentless hunter of great intelligence and stamina, there is also something unhinged about them. A psychopath's charisma that draws you in close one minute only to make you flinch and recoil the next.

Andy nodded approvingly – the nest looked well tended and freshly repaired after winter – and told me there were several other nests in the wood, pretty standard for goshawks. After muscling in on a new patch, they often take over the old nests of other raptors such as buzzards. At other times they build their own, but either way they generally have three or four from which to choose. With a pathological fear of being seen, goshawks frequently nest within the comforting gloom of tall conifers. Douglas firs are a favourite, though occasionally they also go for the open canopies of European larch. When they do, the larch is almost always growing close to dense evergreens that help shield the goshawks' mysterious ways from prying eyes.

I often wonder whether their choice of tree is influenced by what they were raised in as chicks. Or whether it is simply ingrained in the DNA of birds whose ancestors haunted the boreal forests long ago. Whatever the reason, I was relieved that this pair had chosen larch. Even in full leaf these deciduous conifers remain exposed and airy – all the better for filming the nest from an adjacent tree.

Raising my own binoculars, I saw daylight through a lattice of lacy twigs woven around the nest's rim. There was no sign of a bird within, but that was to be expected, since it was still technically winter. Still, spring was fast approaching – one of the reasons we'd chosen to crack on so soon after I'd returned from Kenya. As with most things in nature, timing was critical and any preparation of the site for filming had to be done as soon as the birds had chosen a nest, but before they'd laid eggs. A very narrow window of opportunity.

The larch's bark was a warm chestnut colour, but cold to the touch. Starting with my back to it, I scrutinised every neighbouring tree for possible vantage points from which to set up my camera. Both Andy and I gravitated towards a couple of prominent Douglas firs standing twenty metres away uphill.

Foresters had recently thinned out the wood, leaving the ground littered with offcuts and discarded branches, but this had also opened up flight paths and corridors for the birds. Goshawks love to fly low and fast, skimming the terrain with their powerful chests, before pulling back into steep climbs at the last possible moment – appearing on the nest with no warning. I'm sure this is a conscious effort to keep a low profile, their steep ascents mirroring the vertical tree trunks around them. In any case, the open understorey of the wood was a gift for me also and with a good view of anything going on below me, I'd have a few seconds' warning of their approach.

Joining Andy at the base of the two tall firs, I could see that they loomed over the shorter larch downhill. They should offer a decent view of the nest, but the only way to be totally sure would be to climb them.

Andy pulled on his climbing harness; I stepped into mine, the intrusive clink of carabiners muffled by moss and leaf litter. The firs lacked strong branches low down so the best way to get up them quickly would be to use spurs, as if scaling a telegraph pole. I strapped my climbing spikes to my boots and removed the protective wine corks from their savage steel points. Having passed my safety line around the back of the trunk, I gently placed a razor-sharp point against the thick bark. Stepping up, I felt the blade slide slowly into the cambium as my weight was transferred onto the stiletto. My left foot followed suit and I began to climb, sliding my loop of rope up the back of the tree as I went. The snap of small branches told me

Andy was also off the ground, heading up the fir closest to the nest. The citrus scent of conifer resin filled the air as steel punctured sapwood.

At fifty feet up I was level with the nest, with a clear view down into the cup-like bowl in the centre of its twiggy platform. A nest within a nest, this cup had been carefully lined with fresh green sprigs of fir, a soft cradle awaiting the arrival of a clutch of eggs in a couple of weeks' time. Andy had monitored enough gos nests to know what needed doing for the camera so made short work of any intervening snags and branches, but the exposed view also revealed an awkward branch within the larch itself. Reaching out towards me like a long claw, it obscured the nest and needed to come out. So, leaving a discreet guide rope behind for next time, I abseiled down, gently pushing the ruptured spike wounds back into the bark as I passed. The sub-cambium was flashing pink and I was worried the marks might draw attention to the site. Spring sap would soon be rising and by pressing the soft bark back into the wounds, the gashes would quickly heal.

Climbing a nest tree always feels wrong to me. As if I'm trespassing on sacred ground, which of course I am. Not that it bothers the birds so early in the season, when there is little to tie them to a tree save the time they've invested in repairing the winter-worn nest. This changes once they've laid eggs, of course, but I still felt like an imposter stepping up onto a stage, the surrounding wood a hushed auditorium of watching eyes. Half expecting to be caught red-handed by a returning hawk, I made the climb as quickly as possible, neatly sawing through the long branch at its base. I couldn't resist a quick peek into the nest. Not only had they built in a larch, but aside from the soft cushion of Douglas fir twigs in its centre, they'd also chosen larch for its construction. Large dead branches on the bottom, thinner ones on top, with long whippy twigs woven around the parapet. Larch branches are covered in knobs that interlock, providing rigidity and strength. Such nests are less likely to blow out during winter storms, so they get bigger and bigger each year as new material is added.

Standing on a horizontal limb to the side of the nest, I looked out into the surrounding wood for a goshawk's-eye view. This was the vista the female would have for at least a month as she sat tight on her eggs. From up here the wood became a three-dimensional landscape of dense foliage and distant glimpses. The understorey below was an open colonnade of vertical trunks, but level with the nest the branches closed in and I saw corridors of approach that remained invisible from the ground. A labyrinth of shifting parallax. For a predatory bird able to curl, tuck and swerve through the smallest of gaps, that discreet canopy world would be paradise.

The grey-brown columns of fir trunks led my eye back to the ground where Andy was already down and shouldering his harness. The creeping paranoia of being watched by the resident hawks brought me back to myself, and I descended as quickly and quietly as I could.

The plan for my next visit would be to rig the filming platform and hide. Leaving the wood, I glanced back at the tree. The stage seemed set.

Fast-forward three weeks and it seems now that our filming plans may come to nothing. Andy's voice sounds almost apologetic down the line: 'I've just been over to check and there's no sign of her, James. Probably best if we spend a day trying to find you an alternative. When can you get back down?'

There is no denying that spring is gathering pace. Blossom clouds the pear trees and while winter still clings to the bare branches of oaks, the birches are already in full leaf and the reed beds sway in sunlight that's so much warmer than it was just a few days ago. The goshawks won't wait for us. Two weeks into lockdown, I am still waiting for my official permission to travel and until I receive a letter of authority from Forestry England, I can't go anywhere; nor do I want to.

Thursday 16 April

The low morning sun sets fire to the mist in our valley as I pull onto the deserted road through the village. My official permissions to work, including the all-important licence to film goshawks, came through yesterday, so I'm heading back down to the forest again to meet with Andy. This might be our last chance to find a viable nest before eggs are laid and the goshawk breeding season starts in earnest. The local ravens are already up, rowing across the sky to my right, and I pass our resident buzzard perched like a security camera on her telegraph pole at the T-junction. Even as a silhouette she's instantly recognisable: a massive chocolate-brown bird with a pale crescent moon on her breast. The valley's full of her kind; around twenty pairs. This female and her mate have spent late winter stooping and soaring over the ridge behind us. It won't be long until they're back on their nest, a huge ivy-tangled affair in the crook of an oak in the centre of the local wood.

The M5 northbound is utterly empty. The idea of an empty motorway should be appealing, but it isn't. It's foreboding and unsettling. I'm glad to join the smaller roads winding south towards Warminster. I begin to relax a little; to slow down and enjoy the contours of the valley. I've travelled this route hundreds of times, in every vehicle I've ever owned and at all times of the day and night.

Some of my earliest memories are of sitting in the cab of my dad's Luton van as we drove up from Dorset to sell furniture in Bristol. This is the heart of Wessex, the green country rolling past like a timeline of English history. I've never seen it like this. The landscape shines, but there's no one else here to enjoy it. I alternate between feelings of wonder and uneasiness.

Coming down into the Wylye valley, I come face to face with a muntjac deer. It's standing in the middle of the empty A36, its hog-like back and tucked-in legs making it instantly recognisable as it browses on sprigs of wind-blown leaves strewn across the tarmac. I slow to a crawl and stop 20 feet away in the deserted road. I turn off the engine and watch in silence. He carries on regardless, seemingly unaware that he is dining on what would normally be one of the busiest A-roads in the country. A few minutes later, having crumpled the last of the young tender shoots into his mouth, he saunters back into thick roadside cover, as if making a point of moving on because he's finished, not because I've interrupted him. I've got a lot of time for muntjac, having encountered them often enough in their native rainforests of South East Asia. Love them or loathe them, these descendants from deer-park escapees are here to stay in the British countryside and as I continue on my way, I can't help thinking that some of our issues with their presence here might be because they've not asked permission to take the opportunities we ourselves have offered. For me, this is one of nature's most endearing traits: when all's said and done, it owes us nothing.

A glimpse of Salisbury Cathedral tells me I'm approaching the northern borders of the New Forest. Still dominating the medieval city eight centuries on, it's the perfect consummation of timber and stone. I helped with its restoration once. Climbing past medieval carpenter marks, I'd ascended the iron-like oak scaffolding inside the spire to squeeze out of a small hatch 400 feet above ground. The city below had looked like a model village. Hanging on a rope, I'd chatted to lonely gargoyles while replacing crumbling masonry with crisp new blocks hoisted up from far below. Some of the gargoyles were handsome, albeit in a demonic sort of way. Others were nothing more than a pair of dissolved eyes. But all had been up there, muttering away for centuries. One thing was for sure: they'd borne witness to more than one epidemic in their time. Cold comfort.

Arriving at Andy's house, we keep an unnatural distance from each other, now known the world over as 'social distancing', and drive into the forest in convoy.

Heading up onto the open heathland, we drive past the valley where I once found all six species of native reptile on the same morning. Past the ridge where the wild gladioli grow and round the corner where I'd once disturbed a local trying to cram a freshly killed fallow buck into the back of his hatchback. Frozen in the light of my headlights, he'd had a hand on each of the deer's huge velvet-covered antlers as he tried to wrestle it into the boot. I was just relieved not to have stumbled upon something even more sinister. We stared at each other for a few awkward seconds before he simply shrugged, smiled ruefully and carried on. I'd left him to it, although I couldn't help thinking he'd need something bigger than a Datsun Cherry if he was going to make a habit of nicking dead deer.

I continue to follow Andy's 4×4 down memory lane and soon find myself driving through the tourist-courting village of Burley, with its slightly Gothic, Brothers Grimm facades looming empty and silent – a film set awaiting the cry of action. The local donkeys are clearly enjoying lockdown, presiding over the high street in the morning sunshine.

Back up onto the open heath and ten minutes later Andy pulls over next to a thicket of conifers. The day's nest-hunting has begun and this is the first of the alternative goshawk filming locations he's offered to show me: a huge bulky affair teetering in the top of a small scrappy tree surprisingly close to the road. Not the most picturesque location, and not for the first time that day I regret the fact that our original choice is no longer viable. The site does have promise though, despite the old porn mag lying in the leaf litter below.

I make a mental note of where the filming platform could go and a few minutes later we're back in our vehicles embarking on what will become a nine-hour whistle-stop tour of the forest's best goshawk locations.

We visit at least a dozen or so different territories, spread right across the north, west and central regions. Like most of the keepers, Andy doesn't bother with main roads. Instead we travel with the grain of the woods through a bewildering network of gravel tracks and grassy rides. I used to pride myself on knowing the forest well, but this is a lesson in humility as time and again I have to ask Andy where on earth we are. We visit remote locations that remain free of disturbance all year round, lockdown or no lockdown. We pass beneath the canopies of oaks planted for Nelson, skirt valley mires that still hold isotopes from the Ice Age and discover entire groves of 160-foot-tall redwoods. Such are the haunts of goshawks. It's not surprising that they should seek out these quiet corners. They're hard-wired not to break cover or reveal themselves unless absolutely necessary. And, even then, for only a split second before melting away. Sunshine is no friend of theirs; they seem actively to shun it. The goshawk really is a vampire of a bird.

By mid-afternoon I'm developing what I can only describe as nest blindness. The day is passing in a blur, but Andy is relentless.

He's like some kind of nest-finding machine, finely tuned to detect the nigh-on-invisible traces of goshawk activity. A scrap of down here, a splash of mutes there. A freshly plucked snipe feather half a mile from the nearest heath, or simply an instinctive certainty that this is a place worth investigating. He's known about many of the sites for years, of course, but I hate to think how many hundreds of hours it took him to find them in the first place. Not for the first time, I thank my lucky stars he's willing to help. But I'm also getting nervous. Not one of the nests we've visited is good for filming. Most are obscured behind thick foliage more than ten storeys up, while others are teetering in the tops of spindly trees with no adjacent vantage point from which to film. At a push I could probably make a couple of them work, but it's not looking ideal. By mid-afternoon, out of mounting desperation, I suggest we revisit our original choice. Who knows? It's been a week since Andy last visited and although the majority of nests we've seen are now freckled with the moulted breast feathers of females a fortnight into incubating, it could be that our original choice was simply a late starter. I'm clutching at straws, but Andy humours me and we set off to check it out.

We use a gate to access a normally busy road slicing through the heart of the forest. Regardless of lockdown, I wouldn't usually expect to meet many people deep in the woods, but re-emerging into the deserted public domain comes as a shock. The emptiness is disarming, almost dystopian, although this is more than redressed by the sheer beauty of the place. Emerging leaves hang limp and soft and a promising green haze floats beneath the eaves of the trees on either side. This is one of my favourite times of year. The longed-for return of spring quickens the pulse and I open the window to fill my lungs. We cross a cattle grid, the thump-rumble of our wheels echoing through the vast empty woods.

At any normal time we'd be competing with a convoy of Range Rovers and so I take great pleasure in straddling the middle of the road as we weave our way slowly north towards Mark Ash, one of the most beautiful and renowned woods in all of England. I slow down and lean forward to stare in appreciation at the ancient beeches that I pass. Their familiar forms bring back fond memories of time spent climbing them as a boy, learning the ropes while high in their branches and listening to the whisper of leaves from the comfort of my treetop hammock.

Looking back, I realise how much these childhood experiences shaped the way I now view the world. The climbing skills I learned here have carried me to many rainforests and allowed me to witness things I could barely have dreamt of as a sixteen-year-old. But as incredible as the jungles of the Congo, Amazon and Borneo are, to experience such a beautiful part of England so deserted and free from human disturbance as this seems to me a glimpse of paradise. As I drive, I find myself fantasising about a world without us. How long would it take for nature to reclaim and erase our dubious legacy if we suddenly disappeared? A lot longer than any temporary lockdown, that's for sure. In a country as crowded as the UK, I can't help feeling that even a few short weeks of respite must do the forest *some* good? 'Nature abhors a vacuum', as they say, and I find myself feeling optimistic about what benefits the coming weeks might bring, despite the tragic circumstances that have brought us to this point.

So, reminding myself to make the most of the privileged position I now find myself in, I try to concentrate on the job in hand, namely keeping up with Andy's 4×4 without blowing my tyres.

A short while later we're back at our original nest location. It's almost a month since we pruned out the branches and the air is a little warmer, but once again there's an eerie silence. The serrated tops of firs scratch the blue sky above, but down here the air is chilled and still and there's that same pensive watchfulness that keeps me looking over my shoulder in this little piece of imagined northern wilderness – the home of lynx, wolf and, of course, goshawks.

Goshawks dwell in the corners of time. The moment we blink or turn our heads, they're gone. So, looking out for any flickers of blurred movement, we pick our way slowly forward through the trees. Nothing. I continue uphill to look for the bird's head poking out above the nest but Andy stays where he is, squinting upwards, knowing that she'll be sitting low and tight, and that the best chance of seeing her is from directly below. Sure enough, he beckons me over with a smile. I feel my pulse quicken as I join him. I focus my binoculars on the wispy weave of twigs around the edge of the nest and see the tip of a broad, square tail poking out. 'Like a piece of two-by-one,' whispers Andy. She's lying down with her head hidden, but it doesn't matter because there's no doubt: I'm staring up at a wild female goshawk. And from the size of her tail feathers, she's massive. Not only this, but a scrap of down clinging to a twig tells us that she's started her annual moult and is now sitting on eggs. The incubation period has begun.

It must have started shortly after Andy's last visit on 10 April and a bit of schoolboy arithmetic tells us the first chicks should be hatching around 15 May. Quite late for a gos, or at least compared with the other nests we'd visited today. But it is a huge relief, even more so now that I've seen just how tricky the other sites would be to film. Nevertheless, I tell myself not to get too excited. A lot could happen between now and late June when the chicks would be due to fledge, but it is an excellent start.

The evening drive home is magical, the gibbous moon hanging in a clear sky above the Wiltshire Downs. I don't meet the muntjac again but do slow down to swerve round a boar badger rummaging away in the grass with his not inconsiderable backside hanging out in the road. He barely glances up, despite my open window and a steering-wheel-thumping chorus from the Levellers.

Wednesday 22 April

A week has passed since my nest search with Andy. I've filled the days with lockdown jobs: rebuilt the log store (enjoyable), painted the shed (tedious) and caught up on accounts (downright depressing), but all being well our goshawks are now a dozen days into sitting on eggs. It's almost impossible to be precise about these things unless you have a camera already in the nest, but by my estimation they're about a third of the way through their thirty-five-day incubation period. By now a strong bond should have developed between the female hawk and her eggs and I can easily picture them nestled among soft sprigs of fir, swaddled and protected by the downy breast of their enormous mother.

Goshawks lay between two and five eggs, three or four being the norm. Unlike the beautifully mottled eggs of falcons, those of goshawks are, to my eyes, surprisingly plain and nondescript. A pale-greenish tinge giving way to a dull white a few days after laying. It might help protect them from the attention of egg collectors but does little to hide them from opportunists such as martens, buzzards and crows. I'm sure that a squirrel would also try its luck if it found a clutch unguarded, although a brave predator would quickly become prey should the mother gos arrive back at the wrong moment. Andy and I found a scrap of broken shell beneath one of the goshawk nests we visited last week. It had probably been raided by ravens as there was a pair nesting nearby. Not a lot goes on that these amazing corvids don't notice and they're intelligent enough to wait for the perfect opportunity.

So, between the need to keep the eggs warm and shielded, the female goshawk is pretty much tied to the nest for over a month. Still, she needs to eat and so the male's job is to keep her well supplied with food. He'll make three or four visits a day, delivering the plucked carcass of a pigeon or songbird to an old nest or suitable branch nearby while announcing his arrival with a flurry of loud calls, which draws her off the nest. The male then takes over guard duty, sometimes sitting on the eggs in his mate's absence.

Only the female has the brood patch essential for incubation: an area of warm naked skin on her breast, specifically designed for transferring life-giving heat to her developing clutch. The chicken-like side-to-side movement seen when female birds lower themselves onto eggs helps to part their breast feathers and bring brood patch into direct contact with shell. The male's breast remains insulated by feathers and all he can do is prevent the eggs from cooling too quickly – the equivalent of turning the oven off but keeping the door closed. It's better than suddenly exposing eggs to the chilly air but it's only a temporary solution intended to buy his mate a little time to eat in peace.

While incubating, the female also starts her annual moult. All birds need to replace worn-out feathers, which take a real battering when hunting in dense woodland. The bigger the bird, the longer this takes, so the nesting period is a perfect time for female goshawks to begin the moult. Her primaries are the first to go – those long, stiff, blade-like feathers on the outer edge of her wings. These are dropped in opposite pairs and replacements grow at the amazing rate of a centimetre a day. Even so, regrowth still takes around five months to complete, and may not include the tail feathers, which often have to last a couple of years.

All of this places even more emphasis on the male's ability to provide the female with ample food. Not only is she incubating, but her hunting ability is diminished due to the gaps in her wings as feathers are replaced. Between incubation and moulting, this is a period of low activity for the female. The nest seems to exert an almost magnetic attraction for her that wanes only once the hatched chicks are old enough to be left uncovered. It's this invisible tether that also enables me to rig the site for filming. A certain amount of disturbance is inevitable, but as long as I don't cross the line by pushing things too quickly, the strong bond between female and nest should override her instinct to get away from me.

My plan is to adopt a 'little and often' approach, installing my camera hide in stages. It's a question of gently habituating her to the point where she accepts me as an unwanted but essentially benign neighbour.

All of this is fine in theory, but there are no guarantees. Ultimately, it will depend on what's going on inside the bird's head, and the mind of a goshawk can be an extremely unpredictable and volatile place. Some are skittish; others brazen. Some lie low and stay put; others slope off the nest and melt away the moment anyone sets foot in their wood. The worst-case scenario is a bird fleeing her nest in a blind panic, not only leaving eggs uncovered and exposed, but also potentially damaging them in her haste to get away.

I've been through this same process with many different raptors over the years, but whether it's an eagle, condor or buzzard, installing a filming position opposite an active nest is always nerve-racking. You never really know what you're letting yourself in for until you take that first step. The bird's safety is paramount and always comes first, but things can get tricky sometimes, especially if the adult birds understandably try to drive you away. With these New Forest goshawks, however, my real fear isn't so much being attacked as unwittingly compromising their breeding success. No one wants a failed nest on their conscience.

Thursday 23 April

It rained overnight and the pre-dawn light is sultry as I load the van for the journey south. I hesitate, not really wanting to step out into the real world with all its virus-related threats and insecurity. Lingering in the front garden, I take the measure of the day.

The air is tinged with the heavy scent of lilac and raindrops jewel the newly opened apple blossom. I can't resist burying my nose in it. Not as pungent as the lilac, but just as beautiful and the drops of scented water are refreshing on my tired eyelids.

Still stalling, I let the chickens out of their coop for the day. Officious clucking accompanies their businesslike strut to the feeder, which opens with a harsh metal clang as they stand on the treadle. People think chickens are stupid. I'm not so sure. Slow, maybe; but certainly not stupid. Our two veteran hens, both long since retired from egg-laying duty, haughtily ignore me, as is their wont, and once again I'm struck by how reptilian and prehistoric they look. Stabbing at the feed with mechanical jerks of the head before stalking over to sip water from the hanging dispenser.

A throaty call makes me glance up to where a wood pigeon is perched on the telegraph pole above. I recognise him as one of our garden residents. He too is feeling the pull of spring and is all puffed up to woo the ladies, although I've yet to see him succeed.

Some people think that pigeons are also stupid. But this I know for a fact to be incorrect. The enormous size of this one supports my sentiment. Over the past year he's learned how to open the chicken feeder, timing his visits for when the chickens are in their coop, hopping down to waddle onto the treadle. The lid obligingly opens beneath his weight and he tucks in. The early breakfast shift before the hen house opens belongs solely to him.

His antics cheer me, and even the kids look out for 'El Pijeoto', as they call him. He's certainly the avian Don around here. I have a nasty suspicion he may grow too complacent and fall prey to our local sparrowhawk. Unbeknown to El Pijeoto, this is exactly what happened to his predecessor.

One summer morning we were greeted by the sight of a large female spar mantling the splayed carcass of a very big wood pigeon on the front lawn. She'd obviously hit it hard, feathers everywhere, and my first surreal thought was that the kids had had a pillow fight. The hawk could plainly see the boys and me watching through the window, but the carcass was too heavy for her to fly off with, so turning her back on us she'd spent the next hour systematically taking the pigeon apart – much to the macabre fascination of the lads. Having eaten her fill, she flew off to digest, but returned later that afternoon for a second sitting.

By the time it grew dark she'd eaten the head, the innards, opened a hole in the chest to pluck out the heart and consumed most of the leg meat. Strangely, though, she'd avoided the breast, which to me looked the most enticing. Knowing she wouldn't be back again and that the carcass would otherwise be carried off by a fox, I used my penknife to slice off the two burgundy-coloured fillets and plopped them into a pan of melted butter and garlic. The boys and I had them with eggs, and we couldn't have wished for a better supper. Some of the finest-quality meat you could ever hope to eat, and about as free range and environmentally friendly as it's possible to get. So, when my dad teased me about feeding pigeons, I replied that I'm actually feeding the sparrowhawks and – on occasion – us.

There's something circular and neat about this mini food chain that appeals and makes me think more about the relationship between person and raptor. Falconry has always fascinated me. I even trained up my own bird once: a long-legged buzzard rescued from a Moroccan souk. I eventually returned it to the wild on the Algerian border and although ours had been a brief companionship, involving many basic mistakes on my part, it offered a tantalising glimpse of just how rewarding the bond between human and bird can be.

Falconry is thought to have had its beginnings four thousand years ago in Central Asia from where it travelled west over the centuries to become the iconic sport of medieval nobility. The eleventh-century Bayeux Tapestry shows images of Harold Godwinson (the ill-fated future king of England) with trained birds perched on his fist. It seems he was besotted with falconry, owning one of the largest libraries on the subject in the whole of Europe and the tapestry even shows him carrying a raptor while boarding a boat to visit William the Duke of Normandy in 1064. There's something almost uncanny about this association between hawk-loving Harold, William of Normandy and the New Forest established by him in 1079, shortly after becoming Conqueror. It also intrigues me to think that Harold may have kept a goshawk in his mews, although looking at the embroidered images they look more like falcons or even eagles. Maybe a hawk was too lowly a raptor for an earl to bother with? 'A goshawk for a yeoman', as the old saying goes. Either way, in a time before firearms, what better way to hunt game birds for the pot? And as I drive up the valley again, I think about what our New Forest goshawks would make of El Pijeoto. Very short work, I shouldn't wonder.

Once again, the motorway is eerily quiet, as is the centre of Bath. I stop for a moment to look down on the brown water swirling beneath deserted Cleveland Bridge then drive on out of the city, the roads empty apart from the occasional slab-faced quarry truck hurtling around blind bends. Two hares lope across my path. I watch as they dissolve into the stubble on the edge of a field.

Entering the forest, I pass a group of mothers pushing prams across the tightly cropped grass of a roadside lawn. They stop talking and fix me with angry looks as I slow down to pass. Right now, a van with blacked-out windows is bound to call attention to itself and, not for the last time, I wish I had a more discreet vehicle. Lockdown is causing divisions all over the UK and small local communities are starting to draw in against the outside world.

Despite the stay-at-home message, a steady trickle of campervans, tents and even bushcraft shelters have been cropping up in the forest as people leave the cities in search of a safe, less-crowded place to hole up during lockdown. The forest keepers I meet tell stories of strange folk hanging out in the woods and, in a particularly medieval throwback, there have even been rumours of deer injured by poaching. Keepers can't arrest people, but the sight of an official dark-green 4×4 generally does the trick without the need to call in the police. Even so, there's an unnerving dystopian air to all this as tiny cracks in society begin to show.

Driving down the gravel track towards the goshawk wood, I notice a small brick-red car parked up under the eaves of a spreading oak. I pull up alongside and open the window to chat to its owner, a young bloke in his twenties wearing a camouflaged hoodie.

This is the first time I've met Matt, a local camera assistant, back home from university. He's offered to help us since Covid has put his course on hold and leaning across my van's passenger seat to introduce myself I'm reassured to find a friendly face looking back at me, likewise a whole stack of hand sanitiser, face masks and latex gloves next to him. Once filming starts, Matt will help find locations, lug kit, maintain cameras, listen to me grumble about shots missed and download the footage we *have* managed to capture.

A few minutes later we hear the rumble of a 4×4 on gravel and Matt and I swing into line behind Andy's green Hilux as he slaloms off down the track in a cloud of dust. There's still much confusion in the media over how long the virus lingers on the surfaces we touch, so Andy wears gloves to open the first gate into the woods.

Ten minutes later we pull over to the side of a forestry track running along a ridge directly above the nesting site. The ground is muddy – a dull grey clay flecked with gravel, its surface crumpled and churned by heavy logging vehicles. The plantations to either side have been recently felled as part of routine harvesting. The New Forest is now a national park, but remains a working environment where forestry is very much part of life. The narrow tracks made by my van are dwarfed by the deep, corrugated ruts churned up by tractors and HGV low-loaders. Discarded brash and deadwood litter the broken ground like the aftermath of an artillery strike and cloudy rainwater sits in hollows beneath toppled trees. I catch a glimpse of a roe buck leaping away to my right and a great spotted woodpecker chips away unseen. To our left lies a grassy ride leading towards a line of tall dark conifers. The edge of goshawk territory.

The last time we were here, Andy and I approached uphill from below. This time he's brought us to where we can drop down from above, making it a lot easier to carry in kit.

Matt and Andy chat quietly as I unload the van. Two heavy bags of climbing gear plus the filming platform itself – dents in the metal standing testament to long years of hard use. Made for the BBC *Planet Earth* project back in 2005, it's since been around the world many times and spent countless weeks in almost every jungle imaginable. The last time was January in Costa Rica. Before that: Madagascar; before that: Malaysia. But those lush tropical days now feel a lifetime away; all foreign filming cancelled overnight as lockdown kicked in. Losing so much work has been a kick in the teeth, but at least I'm safely home in my own country. With daily death tolls now rising rapidly, it remains to be seen just how safe the UK really is.

As I wrestle the platform on to my shoulder, Andy grabs a kit bag while Matt hefts the other, and the three of us head into the woods.

Andy chats quietly as we walk in order to avoid a last-minute panic in the bird. Better to warn her we're on our way in. Goshawks are intelligent, calculating creatures with an ability to rationalise in a way that a buzzard, for example, simply can't. We hope her desire to protect her eggs will overrule any impulse to fly. And this can happen only if we give her ample warning.

We keep eyes open for any fleeing shadow but arrive at the base of the platform tree none the wiser. I waste no time hauling up my climbing rope as the clock begins ticking.

The lean of the tree swings me out to the side of the trunk, so I use a lanyard to pull me in close. I climb quickly and quietly. The nest is 70 feet away and I won't know whether the bird's still on it or not until I get level. The only sound is the whisper of a northerly breeze, the clink of climbing kit and my own breathing. My heart is beating faster and it isn't due to exertion, so I'm greatly relieved when I finally glimpse a slate-grey feathered back lying low among the branches of the nest. Long, stiff tail feathers protrude to the right. I can't see her head but have no doubt I'm being watched. Few birds can glower like a goshawk and I know that her eyes will be scrutinising me from beneath their fierce brow ridge.

This is my first proper view of her and even though she's obviously a formidable bird, there's also an air of vulnerability about her. She clearly doesn't like being looked at, but neither does she want to abandon her eggs. She must be feeling very uncomfortable with our forced proximity. This is a critical moment for both of us. How we behave towards each other now will set the precedent for any future relationship. Not that she realises this yet, but we shall be spending a lot of time in each other's company over the coming weeks and the last thing I want to do is ruin things before we've even begun. It takes time to build trust and although I realise that the best I can ever hope for in return is grudging tolerance, she will always have my utmost respect. To help us get off on the right foot from the start, I avert my eyes and do my best to put her at ease by ignoring her as I crack on with the job at hand. I'm pleased to see that for now she seems content to stay put, despite her inherent misgivings.

Looking down, I give Andy the thumbs up, then signal to Matt to begin hoisting the platform. He gives the pulley line a gentle tug and the rope's elasticity bounces the metal frame clear of the ground. It swings out to the side, so I use my boot to deflect the pendulum before it can bash into the tree on the recoil. Keeping my eyes glued to the bird, I feel the taut rope run through my hand as the load is hauled slowly upwards. I signal to Matt to stop, then lock off the platform to let it hang next to me. Watching the bird closely, I give it a few moments to let everything settle.

Strapping the platform to the tree, I keep my movements slow and quiet, glancing nervously towards the nest at every tap of wood on metal. One last look at that motionless feather-plated back, then I abseil back down to the ground as quietly as possible. To have completed such a major task without scaring her away bodes well, but I don't want to push her. Instead, we decide to install the tent-like hide in a few days. It was good to see the female, but her mate, or 'tiercel' as male hawks are known, remains hidden. As we leave the wood, we hear him calling to her from a tall fir overlooking the ridge. He's been watching us the whole time.

Tuesday 28 April

I've been living and dreaming goshawks for weeks, but while there's the opportunity to film in the forest we've decided to add another species into the mix: the rural fox. Not only are they as sharp-sighted as a hawk but they're finely tuned to detect even the faintest of smells. They are also legendarily cunning and wary. And for good reason, given the amount of crap they've had to put up with from us over the centuries. Not that this is all one-way, of course, but I find them entrancing and endlessly fascinating. I love the thought that at least one wild canid still manages to live alongside us. But, for all their familiarity, they can be devilishly tricky to film.

Although the same animal, rural foxes behave like a totally different species from their urban cousins. My first urban fox encounter in a south London garden was a real eye-opener. It sauntered across the grass to snaffle a burnt pork chop before gnawing it in a patch of sun next to the shed. A familiar scene for millions of Brits but having grown up where foxes have been hammered for generations and are supernaturally skittish as a result, it was quite the revelation.

A mate even came downstairs to his kitchen in Brixton to find the back door wide open and an urban fox standing on top of the dining table tucking into leftover curry. They've certainly got style.

Truth be told, I've always wanted an excuse to learn more about them and I realised that I'd never had the opportunity to sit in a hide and film the daily life of a family at a wild forest den. When it came to finding one, Andy's response was exactly as I expected.

'You come to the New Forest – home of some of Europe's most endangered ecosystems – and you want to film one of the world's most common global species?'

I couldn't help but smile as he'd continued.

'They're not wolves, you know! You're not going to get shots of it hunting antelope or anything. All you're going to get is a few fleeting glimpses of a tail disappearing into a patch of brambles. But we'll find you an earth, don't worry.'

It was worth the ribbing.

A month later and I'm in the forest to meet one of the other keepers who has been keeping an eye out for me.

Matt D's beat is centred around the north of the forest, where the mood today is melancholic, wet and brooding. Parking up next to Matt's 4×4 at the end of a gravel track, we immediately attract the attention of a local resident who comes out to see what we're up to. It's now hacking down with rain and he's wearing slippers, pretending to check for post at the five-bar gate while casting us sideways glances. He recognises Matt D and they have a quick chat, which puts his mind at ease. He picks his way back through the puddles to his front door and I picture him returning a rusty old blunderbuss to its position above the fireplace. Still, it must be unnerving to see vehicles parking up right outside your isolated house in the middle of lockdown.

Matt's properly kitted out for the horrendous weather, whereas I can already feel the rain soaking through my torn overtrousers. I vow to join the rest of the world and make a few new online purchases when I get home.

We walk into the wood through the strengthening downpour. Grey running water foams along the compacted soil of the main path, so we step over onto the leaf litter and cut through the trees, the sound of our footfall drowned out by the heavy rain. I'm immediately taken in by the palpable sense of age and mystery of the wood.

Forests of all kinds have always captivated me. So much more than the sum of the individual trees, I see them as entities in their own right. Single interconnected organisms of bewildering beauty, complexity and depth. There is a sublime chaos about ancient woodland that speaks of perfect natural balance, and for me, such places nourish the soul like no other environment. Humans are sensory beings, we all want to feel alive to prove we're not wasting our short time on this planet, and I find the best way to connect with the here and now is to step into trees and give myself over to the wonder, curiosity and joy that they evoke. They help remind me of who I am, where I've come from and where – ultimately – we are all going.

For the past twenty-five years I've made filming in the forest canopy my speciality and whether I'm climbing alongside orangutans in the rainforests of Sumatra or squirrels in the canopy of a local park, I experience the same enticing feelings of being drawn in. Walking through the dark, sodden wood with Matt D, there's that familiar pull of not just wanting but needing to discover what's around the next corner. Always a good sign.

My experience of fox earths is that they're usually located in dense cover: the thick brash of a bramble patch or the low-slung foliage of a young plantation. So I'm taken aback when we arrive in a picture-book open glade. An old beech leans across our path and there, beneath its moss-covered trunk, nestled between gnarled roots, lies a discreet hole. More moss forms a soft cushion on its threshold, adding to the fairy-tale feel. The thought of filming a vixen suckling her young in such an idyllic setting is almost too good to be true, but Matt is confident it's in use and that she's likely down there listening to us now. Vixens will move their cubs on a whim if they feel threatened, so being careful not to flood the area with our scent, we gently retrace our steps. There's no reason to doubt Matt's certainty – it's his job to know where foxes breed – but, just to be sure, we install a trail cam nearby. At the very least, the remote camera will provide useful information about what time of day or night the dog fox is visiting with food.

Driving home, I pull over next to one of the enigmatic Bronze Age burial mounds that pepper the north of the forest. Its summit is crowned with scrub, and for four thousand years it has stood here while the world has ebbed and flowed around it. Entire ages of man have come and gone since it was raised, and I think about the person who lies beneath. What was their name and what had they done to deserve such a timeless memorial? Is it the last resting place of one person or instead the ancestral burial ground for generation after generation of the same tribe?

It's a stark reminder of death and one that taps into my own fears for family and friends right now. It's the uncertainty of the pandemic that plays on my mind most. Realising that hiding from the news won't help this, I force myself to switch on the radio. Twenty-five thousand people have now died in the UK alone. A thousand a day since I last tuned in. As I turn it off again, the news reader's stiff-lipped voice is replaced by the sound of rain on my van's roof. I watch the sombre day bleed into a watery grey as I open my Thermos and look out on to the mound.

Wednesday 29 April

My hide has been up on the goshawk platform for a couple of days now and the finished set-up resembles a *Dr Who*-style Tardis strapped to the side of a tree 50 feet above ground. The fir is strong but thin and I'll need to be careful as I add a heavy tripod, camera, kit bag and large Englishman to the equation. It should be fine as long as the wind doesn't get up.

On top of this, I'm nervous about how the goshawks will react to me filming. Until now I've stayed only a short while to rig, but from this point onwards I'll be up there from dawn to dusk, which at this time of year means about fifteen hours. More than long enough to outstay my welcome.

The plan is to climb up during the ambiguous dimpsy of pre-dawn, but there's no hiding from them, however dark it is. A lot depends on the age-old deceit of having another person walk me in, then leave once I'm settled in the hope that the birds will believe all intruders have departed. It's not seamless when one of us has to climb to access the hide, but it's better than nothing. The birds are everything and no film is worth compromising a nest for; I will be treading softly.

My alarm goes at half past three in the morning. Shower, coffee, porridge. On the road by 4.15. I'm cold. I always get cold when I'm on edge. Nevertheless, yesterday's rain has blown through and there's the faintest glimmer of light over the nearby heath.

Brockenhurst is still sleeping as my headlights illuminate a small shuffling shadow, the epitome of vulnerability. I stop and approach the hedgehog, making sure to note which way it's facing before it tucks and curls. I pick it up and carry it across to the other verge where it remains locked in its prickly ball. As I come into Lyndhurst, a badger trundles ahead of me up the one-way system; through Emery Down, a large dog fox stops to stare. Having satisfied his curiosity, he lopes off, continuing his rounds. He'll have cubs to feed, so I throw a boiled egg out of the window as I pass.

Entering Whitemoor Glade, I slow to let a herd of fallow deer canter across the road. Pale shadows flit through my headlights like ghosts before vanishing into the darkness beyond. We're a month into lockdown and just like the muntjac I met recently, all sorts of animals are now using the roads. A little space goes a long way.

Twenty minutes, ten ponies and a large Highland cow later, I'm rolling slowly down the track to the goshawks.

The eastern horizon is already a few f-stops lighter than the darkness beneath the trees. Matt is waiting at the first gate, dressed in camouflage, the blue flash of his latex gloves incongruous in this rabbit hole of a forest. I pass through into the trees. The dawn chorus has not yet begun, and so the only sound is the slow crunch and pop of gravel under tyres until a tawny owl calls from a large oak on my right and a roe buck steps out to browse the brambles along the fence line. They're one of only two truly native species of deer, the other being red. Roe breed during the height of summer, barely six weeks after giving birth to twin fawns. This is made possible only by delayed implantation and is believed to be an ancient adaptation to life in the wake of our last Ice Age: a time when winters were long and bitter and the growing season short. It's a trait shared by badgers, pine martens and grey seals: an eclectic native menagerie.

I turn off the headlights and crawl forward through the half-light. Newly unfurled bluebells float like grey mist beneath the beeches to my right, while the dark ramparts of the goshawks' conifers loom ahead. I snuff out the engine and take a few seconds to let my hearing adjust. The wood's waking up and Matt and I stand like statues in the half-light to listen.

A robin's piccolo kicks things off with a deceptively simple trickle of notes. A blackbird adds his rich timbre to the refrain. Song thrush follows, as does wren, wood pigeon, then blackcap. Before long the air is filled with the rise and fall of cadences from at least a dozen different species. This exquisite tone-poem is soon lifted to almost Wagnerian heights by the addition of woodpeckers drumming and at least three different cuckoos calling from different points of the compass. Then, at its peak, the distant tremolo of a curlew floats up the valley from the distant heath. If ever there was a sound to whisk you off into a place of windswept expansiveness, it's this call. I stand in rapt appreciation as its refrain echoes through the trees around us.

It's a sublime moment. No hum of traffic. No needy whine of motorbikes and no distant rumble of cattle grids. No sirens and not a plane in the sky. Not a single scrap of noise to link the present moment to the twenty-first century and the closest I could ever hope to come to experiencing the sounds of the forest from a thousand years ago. A once-in-a-lifetime vignette to be savoured and remembered in all its surreal but glorious detail.

The need to get settled in the hide before sunrise pulls me out of my reverie. Shouldering the filming kit, Matt and I begin the slow, cautious walk in. High above, the dark branches of fir slowly close over, choking out what little light had been pushing through.

The birdsong dwindles as we move further in, my breathing loud in my ears as we stalk uphill with the heavy kit. Thick layers of leaf litter that would normally dampen our footfall seem in their warm-spring dryness to yell a loud warning to every creature in the wood. It hasn't rained properly for weeks and I wince at every snap and crackle.

Goshawks rely on acute hearing just as much as vision while hunting in the dim cloisters of dense woodland, so there's no hiding the sound of our approach. Sound draws vision and it's only a matter of time until we're spotted, tracked and scrutinised by laser-sharp eyes.

Animals are surprisingly well tuned in to human body language. Suspicious behaviour is easily detected. We are guilty until proven innocent; animals have learned this the hard way. In any case, goshawks are supreme ambush predators so know every trick in the book. There's simply no deceiving them. Especially when carrying 50 kilograms of camera kit across a bed of dry cornflakes. We've also decided to wear the same clothes each time we visit. I'm pretty sure that wild goshawks learn to recognise individual people and while we may not be able to make them trust us, we can hope to be tolerated.

Arriving at the base of the platform tree, I slowly pull on my climbing harness. Clipping into the rope, I step up to take in the slack, trying to muffle the rub of rope through climbing gear and the scuff of boots on bark. The rope pulls tight, and I feel its nylon stretch under my weight. A glance at the nest shows no sign of movement. All is shadowy and still up there. Placing my trust in the strong bond between bird and egg once again, I continue my cautious climb. A short rope passed around the back of the tree stops me swinging around and keeps the trunk between me and the nest. I resist the temptation to stop and peer as I climb. Raptors hate being looked at, so, keeping my eyes averted like a royal subject, I slowly inch my way upwards.

The dead snags of the understorey are gradually replaced by elegant curtains of tasselled fir as the canopy closes in. Tiny goldcrests flit through the needles and a clatter of wooden wings tells me I've been spotted by a covey of wood pigeon. Arriving next to the platform, I'm now hidden by the hide, so take the opportunity to hang quietly and let the air settle. The climbing rope draws my sight down to where Matt is sitting on a fallen tree, eyes locked onto the nest for any sign of a fleeing bird. A low whistle and he looks up, his face pale in the twilight. He makes the OK sign to let me know there's been no movement and I risk a peek at the nest. Peering through the narrow gap between tree trunk and hide, it takes a few seconds for my eyes to adjust to the dim distance. Eventually I'm rewarded with the now familiar sight of those long tail feathers poking off to the right.

Carefully, ever so carefully, I reach up to unzip the entrance to the canvas hide while pulling the fabric tight to ease the sound. I slip on to the platform. Another low whistle and Matt starts hauling up the camera in its camouflaged bag. This is followed by the tripod head and a small rucksack containing food, water and batteries. It takes me twenty minutes to get set up in the darkness of my new home. One final whistle to let Matt know he can leave, and I hear him walking off through the trees, humming softly to draw the bird's attention in the hope she thinks we've both gone.

It's now half past five and I gingerly open the hide's draw cord to allow the front of my lens to peek out.

The image on my camera's screen is flat and underexposed, but I can clearly see the bird's steel-plated back. I watch the gentle rise and fall of her breathing and allow myself to relax a little while waiting for sunrise.

It is almost May, but the wind hasn't yet swung round to the south. I can still smell the breath of winter on the north-easterly sighing through the conifers. A cuckoo still calls, but the dawn chorus has ebbed away, and the wood is now almost silent. In this strange, difficult, frightening year, the morning air is hollow, clean, fresh. Timeless and ageless and pure.

It is now light enough for me to take a closer look at the nest through the tight end of my telephoto lens. The female is still flat out, head low while warming the eggs hidden beneath her. Caught on the knobbly larch twigs, a few wisps of moulted down sway gently in the chilly breeze.

Somewhere out there beyond the wood, the sun is rising. A few minutes later and the first tentative rays reach through a gap in the canopy, an almost horizontal beam of rose-gold creeping down the glowing bark.

I often wonder why nests are located as they are. These goshawks have chosen a fairly nondescript larch, one of several dozen within this wood alone, and yet I assume this is not random. Until you spend time watching a nest it's often difficult, if not impossible, to appreciate why they've been built in that exact place. It may be sheer coincidence, but a few moments later the shadows are chased off the nest by a direct beam of sunlight lancing down to illuminate it like the centre of a stage. The surrounding trees shrink back into shadow, but the nest itself glows. By the time I've reeled off a handful of shots, the moment's passed and the spotlight has continued its descent as the sun rises higher. By now there are several patches of light on the neighbouring trees too and rose-gold becomes pale yellow. Goshawks are early breeders and the nights are still cold, so is it too fanciful to think the nest has been deliberately positioned to take full advantage of every ounce of early spring sunshine? I've filmed birds of paradise pecking holes in leaves to direct sunlight onto their dark jungle display courts, so why couldn't a raptor as intelligent as a goshawk make similarly complex decisions?

A loud *kek-kek-kek* from the trees to my right and I instinctively press record on the camera just in time to film the female goshawk's response to her mate's call: a razor-sharp glance in my direction followed by a vertical head-first plunge off the nest down into the comforting gloom of the woods below. She's gone in an instant of power and grace.

But as impressive as she most certainly is, nothing could have prepared me for my first sight of the male. He appears on the nest, utterly silent, as if he's just stepped through from another dimension. I realise that I'm holding my breath as I scrutinise him closely through the long lens.

He's a totally different creature from the female. Much lighter in build, barely half her size, it seems. There's a restless wild beauty to him that speaks of frost, piercing winds and snow-dampened forests. There's also something shadowy, even blurred, that suggests transience, as if he dwells on the wavering edge of visible light. Where she is all about brooding power, he appears almost fey in the true sense of the word. As if he doesn't belong to this world and might evaporate any moment. It's an enchanting but inexplicable effect. This otherworldly aura is further enhanced by two pale stripes that encircle his dark-grey head like a crown. But the most arresting feature of all is his eyes. Whereas the female's are a brilliant, piercing orange, his irises are the colour of rubies or smouldering coals. A red so deep in colour it's impossible to tell where his irises stop and his pupils begin.

As he stands stock-still on the edge of the nest where he landed, only now do I notice his puffed-out tail coverts flashing white like cotton wool. They draw the eye in the low light, and I wonder if he uses them to warn the female of his approach as he flies towards her through the dark firs. Living your life in mortal fear of being attacked or even killed by your larger mate is the goshawk way. Reverse sexual dimorphism is very common in raptors and it wouldn't do to be mistaken for an intruder by the larger female. I can't help thinking that this vivid splash of white might even act like a flag of truce.

Still poised on the edge of the nest, he cocks his head to peer down at the eggs. I can't see them, but he seems fascinated by the sight – as if he's never had a proper look before and doesn't really know what they are. A few heartbeats later, he steps tentatively forward, taking great care over where he places his feet. Leaning forward, he lowers himself gently onto the hidden eggs. I can see from his bulging crop that he's recently eaten and for a moment it looks as if he might doze, but then he stands up to stare down at the eggs once more. This time he shuffles round to the left, lowers himself down again and is just getting comfortable when he suddenly cocks his head in alarm, leaps up and barrel rolls off to the side. I get a fleeting glimpse of his wings unfurling to catch him as he falls, and a split second later the female appears where he was sitting. Her huge yellow feet tipped with long black talons seem to land right on the eggs and it's a mystery to me how she avoids breaking them. This is my first full view of her and as I move the camera in my eagerness to get a shot, her head snaps round and I'm staring straight into a pair of eyes burning with shocking intensity. The contrast between the two birds is hard to overstate. The female seems to radiate strength, a real bruiser of a bird with an immensely powerful, deeply muscled chest, strong back, and long, stiff tail feathers. Not a puffed-up tail covert in sight, I note. Then there's her colour. The male's charcoal-grey plumage had been delicately trimmed with sharp lines, as if adorned with silver filigree. The female's plumage instead carries a hint of brown, with less crisply defined eyebrows, or superciliary stripes.

She continues to hold my gaze for a couple of minutes, as if weighing up her options, then looks down at the eggs, preens a fleck of down from her breast and carefully, very carefully, sits down. A moment later she's up again, turning the eggs gently with her massive beak, before finally getting settled.

It's now midday and she hasn't stirred for several hours. Her head is tilted forward, and she seems to be sleeping. The whole wood is all too quiet. Even the goldcrests have stopped their incessant squeaking from the branches around me. The only sign of time passing is the sunlight creeping down tree stems. I haven't seen or heard the male for hours and am beginning to wonder whether I'd ever seen him at all.

Mid-afternoon. Nothing more has happened on the nest and it's becoming very cold. I've got a rug draped over my knees and have put on another layer of clothes, including waterproof trousers, hooded top and body warmer. Sitting still for hours on end, the chill creeps into your bones and I'm now warming my hands on the hot air expelled from the camera's exhaust fans. It's a different experience from the last time I sat in this hide a few months ago in Costa Rica, that's for sure.

The wind's now blowing hard, setting all the surrounding tree stems asway like a fleet of yachts at anchor. It isn't a pleasant sensation being buffeted like this, at the mercy of the elements and stuck inside a tent, horizon hidden. Keeping the camera steady is also very difficult, like standing on deck trying to film boat-to-boat in a rough swell. I move one way, while 70 feet away the nest tree moves the other. I've set the camera to film in slow motion, to take the edge off the worst of it.

I'm also acutely aware of the extra weight I've placed in the

44

tree: as strong as Douglas firs are, there are no guarantees. Trees can cope with a lot but there are limits and the effects of strong gusts on overloaded timber can be scary, so I may need to rig guy lines to help limit sway and reduce stress. I'm hoping that these will act like shrouds on a mast. They'll certainly make me feel better. The thought of a tree snapping while I'm trapped inside a tent 50 feet up doesn't make for a relaxing working environment.

However, as tricky as it is to stabilise the camera in this wind, it's nothing compared with remaining balanced while peeing into a plastic bottle. From the sublime to the frankly ridiculous.

Saturday 2 May

3.30 a.m. Wake up. It's my second filming day and I'm hoping the new guy lines I installed yesterday will work. From a distance the tree now looks like a maypole but happily the female sat tight on the eggs as I worked, sloping off into the wider wood only as I finished and the male called from the trees with food.

I turn on the bedside lamp to prevent me from going back to sleep, grab a shower, then stumble downstairs to the kitchen of my self-catered cottage in the south of the forest. I pull down the window's blind to hide the inky blackness outside. I've always hated being inside a lit room when I can't see what's out there. Some sort of remnant paranoia. It's the same primeval feeling I get at dusk in a rainforest as shadows rise and the safety of the canopy above seems to take on extra appeal.

Coffee. Mobile phone propped against the toaster so I can watch the latest weather report, looking for those little arrows that show anticipated wind speeds and direction. It's the gusts that count. Anything above 20 mph will be pushing it, even with the guys in place. Still, if you don't like what the BBC says, then there's always another weather site. Just keep trawling until you find a report that suits you, depending on how you want the day to go.

4 a.m. Load the van. It's a faff in the dark, but the rented camera is worth £80,000, so it doesn't do to leave it in a vehicle overnight – even in a place as sleepy as this. I close the boot as quietly as possible, not wanting to disturb the muted air. I can smell the tang of salt from the Solent a few miles to the south.

4.15 a.m. Wait for Matt. My accommodation is next to his home, so I look for the glow of his headlights as he pulls out of his drive and fall in behind. Matt's car has disappeared around the bend as I approach the railway bridge on the outskirts of Brockenhurst. It's pitch-dark beneath it but I catch the flick of a tail in my headlights. I brake hard, coming to a stop twenty feet from a small herd of ponies with dark coats and no reflective collars.

Many years ago, I was first on the scene at a collision between a pony and car up in the north of the forest on a long, dark section of road that cuts across heathland at a place called Godshill. The driver had been heading home after an evening shift and was lucky to escape with barely a scratch, but his car was a write-off. The pony was lying on the road in shock. Blood everywhere, nostrils flaring, eyes rolling. The look it gave me as I knelt to stroke its neck comes back to me time and again: shock, confusion, terror. As its eyes clouded over, I remember thinking how surprisingly long and human-like its eyelashes were. I couldn't bring myself to be angry with the driver. He was young and utterly inconsolable. The front of his small car was completely caved in where he'd hit the pony's legs. The windscreen was a concave mess of shattered glass where the heavy body had hit then rolled back down the crushed bonnet. How the driver hadn't been killed was a mystery.

This was in the days before mobile phones, but eventually another driver stopped before heading on to call an agister. I covered the pony's head with a jacket to hide the reproach in its now lifeless, congealing eyes. Half an hour later a Land Rover arrived, and I left the driver to explain things to the hard-faced forest official. I remember noticing the length of the skid marks on the road as I drove off; the speed he must have been going on that forest road.

Brockenhurst is lamp-lit and silent. We're now five weeks into lockdown and at a local football club, a forlorn sign: 'All fixtures cancelled for the foreseeable future.'

Entering Lyndhurst, the deserted road is strewn with rubbish ripped out of bin bags by foxes and badgers. The local cats also know a good thing when they see it and I'm greeted by the sight of all three animals as they muzzle through the debris, like the aftermath of a Mardi Gras.

Entering Emery Down, I encounter a pregnant fallow doe standing in the middle of the road beneath the pale light of a street lamp, like a subtly lit statue in a museum. She watches me for an instant before stalking slowly into the trees on the left.

Arriving at the first forest gate, I open the van door and birdsong floods in. A blackcap, wren and song thrush vie for air-wave dominance. The bluebells are out in full bloom beneath newly unfurled beech leaves and the wood is brimming with early-morning life as we drive on. Blackbirds make last-second dashes across the track in front of me and robins bob and display from the top of brushwood piles left by foresters. Having felled and removed the trunks, they've stacked the discarded branches and brash into large round cones. In the low light of dawn, these neat, hut-like domes are scattered through the woodland like a Mesolithic settlement.

All the scene lacks is the smell of woodsmoke and the barking of hunting dogs.

4.55 a.m. Matt and I stand at the base of the platform tree. We're still thirty-five minutes away from sunrise, but it's light enough to see the outline of the female on the nest. She sinks visibly lower as she realises she's being watched. A male wren sings from his favourite perch on the roots of a fallen fir. Sacred to the druids and firmly entrenched within folklore by Aesop, round here this plucky little 'King of All Birds' was once called 'cuttran', or simply 'cutty', meaning small. Flitting back down into the bracken, he continues his morning rounds while I ready myself to ascend.

It's raining by the time I'm settled. The sun rises behind a murk of grey cloud and the wood drips as everything becomes saturated by slow, steady sheets of drizzle. Water begins pooling on the flat top of my hide a few inches above my head, so I wedge my climbing helmet between ridge pole and canvas to increase the pitch of roof and aid run-off. My hide hasn't been waterproofed for years and every time it rains I vow that I'll do it. But then it stops, and I forget. But for now, the rain has yet to soak through and I sit in the gloom listening to the crackle and pop of water on leaves.

These are some of my favourite moments in a wood. A little rain goes a long way to creating atmosphere and any honest documentary about British wildlife has to include an occasional wet day. In the words of Alfred Wainwright, 'There's no such thing as bad weather, only unsuitable clothing.' (He should know, having spent most of his life in the Lake District.) There are many different types of rain, but persistent saturating rain from the south is known locally as 'woodfidley' rain. It's a great term, evoking images of fidgety trickles running up sleeves and down the backs of necks. This is a rain that can also saturate colour and soften the landscape. Overcast skies help diffuse light and reveal detail. The woodland floor far below now glows green as powder-dry moss and brittle crimp-edged ferns are brought back to life.

The female is still sitting low, occasionally shaking her head to dispel water. The young spiky larch needles above her carry tiny silver baubles, and droplets lie scattered on her dark back like beads of mercury.

She doesn't look happy. To be honest, goshawks very rarely do, but today she seems particularly over it and I'm not surprised. The fidley rain is clearly doing what it does best and although her feathers protect her from the worst of it, I suspect it's seeping underneath too, forcing her to sit all the more tightly on those precious eggs. On the other hand, she probably hasn't been able to take a proper bath for days and this is the first rain we've had for ages, so maybe it's more welcome than it looks.

Raptors like to bathe regularly. Like all birds, their lives depend on keeping their feathers in good condition. So dirt, blood, faeces and all the rest of it needs to be washed off regularly. These goshawks will have a favourite shallow or pool where they'll spend a few minutes splashing around before retiring to dry out in some high place. The post-bathing preen is just as important as the bath itself. All flying birds have a uropygial gland near the base of their tail: a little pimple hidden beneath their feathers that secretes a waxy concoction used to keep their plumage and talons in good condition. The oil is transferred to each feather in turn by the beak. The feather fussing and nibbling that birds do after bathing is all part of this. There's no point oiling dirty feathers, but neither can a nest-bound goshawk bathe as regularly as she'd like. When she stands she reveals dirty-grey coverts and belly feathers.

In medieval times, the uropygial gland was also believed by some to contain poisons that helped birds of prey kill more efficiently when transferred to talons during preening – a notion first mentioned in the thirteenth century by Holy Roman Emperor Frederick II: 'Talons of birds of prey, owing to the noxious character of this oil, inflict more deadly injuries upon and bring about quicker death of their quarry because the wounds they make are toxic.'

Totally wrong of course, but the idea of a venomous goshawk would be enough to give any woodland creature the screaming abdabs, I'm sure.

Being tied to the nest for well over a month must be an incredibly frustrating thing for such an active hunter as a goshawk. I wonder whether shifting hormones help her cope, but I dare say that relying on her mate for scraps of pigeon quickly wears thin. I have absolutely no doubt that when she's finally released from the shackles of incubation, she'll take great pleasure in returning to the fray. Millions of years of evolution have sharpened, honed and focused these birds into the perfect predators we see today.

Predators obviously need to be physically adept, but they also need to be smart with it and I don't think it's coincidence that some of the planet's most astute hunters evolved among trees. Forests require complex survival strategies, including intricate mental maps, acute spatial awareness, fast reactions and high levels of problem-solving. Goshawks have these attributes in spades, plus a shape-shifting body profile that allows them to navigate three-dimensional labyrinths at speed. They are hard-wired to do what they do best and being confined to a nest for five weeks must be tough. In the immortal words of Dr Alan Grant, 'T-Rex doesn't want to be fed, he wants to hunt. Can't just suppress 65 million years of gut instinct.' Watching a goshawk feeding on a kill is an exciting reminder of the proven truth that dinosaurs are still with us today.

Birds are direct descendants of bipedal, predatory dinosaurs called theropods, which included not only the Tyrannosaurus rex but velociraptors too. Theropods were a diverse group characterised by hollow bones and, in some instances, feathers. Fossil evidence shows that feathers were evolving in theropods millions of years before the arrival of Archaeopteryx, traditionally the earliest bird. I remember being appalled by the concept of a feathered Deinonychus when this was first announced but having spent thousands of hours filming raptors since then, Grant's prophetic words ring true: 'Bet you'll never look at birds the same way again.'

I also believe that goshawks are capable of holding a grudge – or at least biding their time until an opportunity to act presents itself. I once watched a squirrel climb on to a goshawk nest while foraging for seed cones. It either didn't realise or care that the female goshawk was sitting on eggs, scowling at it from barely inches away. Three weeks later, that gos was free to hunt for herself once more. Seeing the same rodent approach, she immediately slipped off the nest, circled round and ambushed it from behind, pinning it to the tree with such force that the thump echoed through the wood. She then spent the next hour feeding said squirrel to her three hungry chicks, while another – having presumably witnessed the attack – huffed angrily at her from a safe distance.

The rain continues and I don't see the male bird until well into the afternoon. The gunfire rattle of a wren's alarm call warns me of his approach. I've learned to listen out for the *rat-a-tat* of the cutty. He's the only bird plucky enough to raise the alarm when all others remain silent. I guess he feels secure in his tangled heap of bracken or knows he's too small for a goshawk to bother with. Sure enough, a few seconds later the male gos slides into view carrying a rook squab. I know it's a baby rook, because the entire rookery is in pursuit and spends the next half hour swirling above the wood in an angry storm of harsh cries. None dare to fly beneath the canopy though. By 6 p.m. a thin veil of mist is drifting between the trees and the light plummets.

Wednesday 6 May

At face value, the New Forest's name is misleading, verging on false advertising. A Geordie mate put it best: 'It doesn't look very new to me, like.' It's an obvious anachronism, harking back to the forest's creation by William the Conqueror in 1079. The origins of the word 'forest' are not quite as straightforward. Today we think of the word as being specific to extensive tree cover, but a thousand years ago the term 'foresta' denoted any kind of land formally set aside as a royal hunting ground. It is rooted in 'foris', meaning 'outside' – in this case, beyond or outside common law; a designated preserve for the beasts of the chase that lay under the direct control and owner-ship of the crown.

The name Nova Foresta itself first appears in the Domesday Book of 1086 and the whole concept seems to have enjoyed a less-than-favourable reception from the start. The problem of course was that William's new hunting reserve was forcefully imposed on a region already steeped in tradition and holding common rights of use such as woodcutting, hunting and livestock grazing that had been in place since time immemorial. Although many of the forest's ancient customs had previously been ratified by King Canute, who had also hunted the area, William's new edict rode roughshod over all of this and significantly curtailed what could and could not be done on this communal land, even by the recently supplanted Saxon aristocracy.

As someone once said, 'Englishmen hate an arbitrary power as they hate the devil', and it's hardly surprising that these Norman land reforms were very unpopular with the local populace, who were often left with 'nothing but their eyes to weep with', as one chronicler described it. The Norman kings don't appear to have placed much emphasis on public relations and I suspect that our nation's inherent sympathy for the underdog can be traced back to the Norman Conquest, if not the Roman invasion a thousand years earlier still. From Robin Hood to William Wallace, we love a good rebel.

In its original form, the name New Forest held these damaging and damaged associations, denoting something entirely different from what it's generally taken to mean today. And although it does indeed contain a lot of trees, it also contains just as much open heath, grassland and bog. It is a constantly shifting mosaic, in fact, and therein lies its strength and importance as an area of extremely high biodiversity and ecological significance. Each habitat contains its own ecosystems and although many species move freely between, there are also many that are obliged to stay put due to their very particular needs. Some species are so closely associated with their niche that they seem to embody the very essence of a place. Just as goshawks and tawny owls might epitomise densely wooded cover, the open heathlands are closely associated with birds such as the stonechat and nightjar and reptiles such as the sand lizard and smooth snake. For me, the lowland heaths of the New Forest have always meant one bird above all others: the Dartford warbler.

I only ever saw one Dartford in the whole time I was growing up here. I knew they were in the area but, try as I might, they eluded me. They are a charismatic little bird, but one whose fortunes have always been closely entwined with conservation issues and our nation's complicated relationship with land development and the climate crisis.

First described in 1773 from specimens collected on Bexley Heath, near Dartford in Kent, it is our only resident warbler. It is a Mediterranean species really, and the British population lives at the northernmost limit of its range, but whereas other warblers, such as sedge, wood, willow and even blackcap, usually migrate south to warmer climes in winter, the loyal, plucky little Dartford stays put. This decision to brave out the British cold has cost it dearly over the years and very nearly proved its undoing. A series of severe winters in the nineteenth century knocked it back considerably, but numbers slowly built up again until the early 1960s when Britain was hit by two exceedingly harsh winters in succession. This took our national breeding population down from an already marginal 450 pairs to a seemingly doomed population of just 11.

The winter of 1963 was the coldest on record since 1740, with temperatures plummeting to −22°C and snow cover lasting from December to March. Feathers are good insulators, but there are limits and it's no evolutionary mistake that animals inhabiting arctic environments tend to have large bodies to help minimise heat radiation. How many of our smaller birds manage to survive even a standard British winter is beyond me, but for an insect-eating warbler weighing less than ten grams, a three-month period of deep snow, clear skies and sub-zero temperatures must have been apocalyptic.

To compound such vagaries, the Dartford's preferred habitat – the gorse and heather of acidic lowland heath – is one of Europe's most threatened and dwindling habitats. The New Forest contains around 10,000 hectares – Europe's largest surviving tract, in fact. Overall, Europe has lost almost 90 per cent of its lowland heath in the past 200 years and even though the UK still holds a fifth of what remains, it is severely fragmented, frequently threatened and a shadow of what it once was.

In a way, heathland has always had a bit of an image problem. Occupying impoverished, sandy and acidic podzol, it was never much use for agriculture and was often regarded as a kind of hinterland or 'lark lees' (land fit for only birds to live in), where anything went. It was also the traditional haunt of those existing on the fringes of society: Romanies, smugglers and landless peasants scratching a living from cutting furze and working the meagre soils. A barren landscape ready for development or commercial afforestation in a kind of waste-not-want-not mentality that began with the eighteenth-century Enclosure Acts and reached its height during the mid-twentieth century. The post-war years saw attempts to build housing on New Forest heathland, but this was diverted, although Canford Heath, 20 miles away to the west, wasn't so lucky and ended up taking the fall. Once the largest tract of pristine heath in Europe, Canford is now home to one of its largest housing estates. I can still remember the raging fires that swept across it when I was a boy. Columns of black smoke rising into the blue sky as 'accidental' burnings paved the way for development.

On top of all this, open heath was also feared as the haunt of capricious if not downright malevolent spirits – hardly surprising for such an austere landscape dotted with mysterious prehistoric burial mounds and shape-shifting wind-torn trees. Shakespeare was tapping into pre-existing themes when he chose a heath as the setting for the meeting between Macbeth and the three witches, while Germanic folklore alludes to its dark magic by filling it with dragons.

For me, it's Thomas Hardy who most successfully captures our often ambivalent relationship with this habitat. His novel *The Return of the Native* is entirely set on Egdon, a fictional heath, near to Canford over the border in Dorset: 'The place became full of a watchful intentness now; for when other things sank brooding to sleep the heath appeared slowly to awake and listen.'

Hardy's Egdon was a character in its own right. A landscape with a 'lonely face, suggesting tragical possibilities', and although Egdon never existed as one single entity, it was based on a very real network of heaths that once stretched along much of the south coast from Cornwall to Kent. Essentially, the forest's remaining pockets are remnants of this same landscape, and while the smugglers are gone and its wildness may have been tamed over the years, its atmosphere has in no way diminished in power. I've spent many nights out on the heath and believe it has an elemental aura entirely of its own.

Many years ago, a friend and I lay in the heather gazing up at the stars. It was a clear, black night with no moon. One of those perfectly still nights when the heavens eavesdrop on the banal musings of mortals. We'd just shared a few pints when I was hit by overwhelming vertigo – nothing unusual there, I guess – but it was as if the surface of the heath had flipped upside-down and we were now staring down into the fathomless abyss of space from above, about to freefall into the beyond. A joyous moment of expanse, as if noticing the stars for the first time, which was quickly followed by the crushing realisation of how fragile, inconsequential and tenuous life on our planet truly is. Like an enormous parabolic reflector, the open heath amplified these revelations to the point where I felt utterly naked and exposed. Crawling back under the cover of trees, I stretched out in the leaf litter and fell asleep beneath the comforting shelter of an oak. My feelings for the heath have remained ambivalent ever since. I find it beautiful and intriguing, unsettling and intimidating, but at all times highly charged.

From a naturalist's point of view, it can appear austere and devoid of life, but peer a little more closely and you'll see that it's actually teeming. You need only know where to look and when. So, in an attempt to sing the praises of an often overlooked habitat, I've decided to revisit the challenge of filming Dartfords.

Since those harsh winters of the 1960s the Dartford warbler has steadily increased in numbers to a fairly secure national population of around 3,200 breeding pairs, and whereas in the past the New Forest has been home to between 400 and 500 of these, Andy tells me there are currently around 300 in the area, which – though worryingly small – still represents the UK's largest breeding population. I've followed their story and learned to love them from afar, but thanks to Matt – who's been watching them on his local patch for years – tomorrow I'm hoping finally to film one.

Thursday 7 May

True to form, the heath is dark and saturnine as we walk out in the early morning. I'd be disappointed if it wasn't. The brooding atmosphere is lightened by the scent of coconut as we push through the flowering gorse and, risking a face full of spikes, I pause to smell one of the tiny yellow flowers. Shaped like a hooded face, the delicate petals bring a welcome splash of colour to the pre-dawn grey and smell divine. Bizarrely incongruous but exceedingly welcome. I inhale deeply, filling my lungs with the memory of sunshine, palm trees and rum: a barrage of imagery that's utterly at odds with our setting.

Following a narrow, flint-strewn track, we snake through the bushes and emerge on the crest of a heather-clad spur that rolls down into a wide mire of pale tussocks and low mist. The full moon hangs over gaunt stems of burnt gorse behind us, while an orange glow spreads along the horizon to the east. Matt sets up the tripod and I click the camera into place just as the sun's enormous rim breaks the horizon. Black silhouettes of trees ripple against liquid red as it swiftly rises. Colour flows into the landscape and, as if on cue, we hear the first skylark from somewhere high above. Within a few short minutes the sun has pulled clear of the horizon and solidified into a ball too bright to look at or even film. Two ravens fly overhead and a flash of white draws my eye to a distant lapwing. Stonechats stand sentinel on gorse and I hear the soft calls of meadow pipits somewhere behind me.

Matt's confident that this is a good spot for Dartford, so holding still we listen out for the male's low-key song. Sure enough, within a few minutes we hear it coming from low down in the gorse to our right. A confused churring of jumbled and split notes. A bit like the song of a robin played back triple speed. Scratchy and over too soon, it's not the most beautiful of melodies, especially compared with the skylarks singing above us, but it seems to suit the bird's humble profile perfectly. A moment later we're rewarded with our first glimpse as he breaks cover, cocking his long tail to balance on a sprig of gorse before dropping back down into heather below. I grab a shot, but it's fleeting and only leaves me wanting more.

This pretty much sets the tone for the next hour as I stand in one spot, frantically panning the camera from bush to bush, getting it focused just in time to catch the tiny rascal disappearing back into dense cover. On the plus side, he doesn't seem fazed by us in the slightest and is soon joined by the female, his mate. They are both foraging for insects, flitting from branch to branch as they collect delicate beakfuls of tiny wings and legs. They keep returning to the base of the same clump of gorse 50 feet away and it's clear they have a nest down there. Their skulking behaviour reminds me of their delightful local name – furze wren – an old-fashioned label that tells you much about the bird: where it's found, what it looks like and how it behaves. In this instance: furze (the West Country name for gorse); small with a cocked tail like a wren's and usually seen foraging in dense cover low to the ground, again just like the cutty.

The only way to film birds like this is to stick to them like glue through the lens, hoping that you're eventually given a break. So I persevere. It's not the most satisfying way to build a film sequence, but all I really need are a few shots to show the bird's place in the heathland habitat. Lucky really because, in the space of three hours, that's all I manage to get.

Finally, much to my relief, the male breaks cover to perch on a twig and sing. He's in the open for only twenty seconds, but it's enough to get my first proper look at him. His plumage is discreet, but immaculate in a country-squire kind of way, the rusty brick-red of his breast contrasting perfectly with the gun-metal grey of his head and back. There's a pale patch on his belly, looking for all the world like a shirt poking out from beneath a waistcoat, and his legs are dapper orange. The beak is particularly wren-like, and his throat is flecked with white, like a cravat. His slate-coloured crest rises and falls as he sings, and his long, ramrod tail is raised to help him balance. The most striking features of all are his eyes. Irises the same fired-clay orange as his breast and ringed with bright vermilion, drawing you in and holding quiet attention. Appropriately enough, my overall impression is of an understated but intriguing character in a Thomas Hardy novel. Even the ochre of his breast seems reminiscent of Diggory Venn's reddle-stained skin in *The Return of the Native*.

A few seconds later, he's gone. Flitting back down into the heather to rejoin his mate in their ceaseless foraging. It's comforting to know that down there somewhere, among the stems of heather and furze, sits a cupped nest of dried grass containing three or four hungry mouths. Dartfords often raise two broods, sometimes even three, and this is probably the secret of their ability to bounce back despite all odds. That and a whole lot of dedicated conservation from the likes of the RSPB and Forestry England. Hopefully, this is one native that's now returned for good, although it's worth remembering that, as a habitat, heathland will exist in the long term only if it's grazed by livestock and properly managed.

Although lowland heath is a wild habitat that first appeared in the wake of the last Ice Age, if left to its own devices it quickly becomes woodland as heather and gorse give way to birch and pine in natural succession. In fact, lowland heath wouldn't exist at all if our ancestors hadn't recleared tree cover during the Neolithic allowing the heath to recolonise. It's only the ongoing maintenance of these cleared areas that has enabled the biome to survive to this day. Such clearance can be done by blade or fire, but burning has the added advantage of removing dead organic material that would otherwise rot down and enrich the earth to the point where it would alter the free-draining sandy soil loved by heather and gorse. Like the plants of many other fire-based ecosystems, gorse contains highly flammable compounds designed to encourage burning. It also has seedpods that open in heat, allowing their seeds to germinate in newly cleared areas after the fire has passed and has an incredible ability to resprout from the roots of burnt stems. All in all, furze is a phenomenally hardy shrub.

So lowland heath needs to burn once in a while, but it has to be done by people who really know what they're doing on a regular rotation that helps avoid the intense destruction caused by overdue burning of mature scrub. Such controlled burning normally takes place early in the year, so that the underlying peat is still damp from winter and able to protect the seed bank. This dampness also prevents the peat itself from catching fire. Apart from the huge release of carbon this would cause, a fire that goes underground can be almost impossible to put out, as demonstrated by the rampant burning of drained peat swamps in Indonesia. In February, most British reptiles will still be hibernating below ground and birds haven't yet started to nest. As Andy tells me, 'Generally speaking, we try not to burn too much of the old heather and gorse areas in any one year, so we're not impacting too high on Dartford warblers.' While some are inevitably displaced, the species isn't at saturation in the forest, so territories can be held and re-established on rotation too.

It's also important to stress that in the forest these small-scale burnings are done in such a way as to maintain different stages of habitat growth in favour of diverse wildlife, not for the sake of just one species of game bird, as with the controversial large-scale burning of upland grouse moors in northern Britain.

Whether or not this arrested development should still be perpetuated is part of a much wider discussion on the concept of shifting baselines and the exact nature of what constitutes a natural environment, but for me lowland heaths are very much a part of the rich tapestry of our island's landscape and although it may seem counter-intuitive to burn an environment in order to protect it, if left to natural succession the acidic heath biome would soon disappear.

The landscape this morning has been devoid of people; the only sounds birdsong and the breeze rasping through the gorse. The road to Brockenhurst, normally so busy, is empty and seems to have been swallowed up by the rising heath: the revolving world of people appears to have ground to a halt. Happy with what we've filmed, we pack up, shoulder the camera and tripod, and walk back along one of the flinty tracks criss-crossing the heath.

Cresting a ridge, we're suddenly confronted by the excited barking of a dog and the distressed calls of a bird beside itself with angst. In a textbook distraction display, a lapwing is flying low over the heather, trying to lure the dog away from its hidden nest. The dog – some kind of saluki – bounds after it but seems torn between trying to catch the bird and returning to investigate the nest. A second bird joins in and the distraction ploy turns into full-scale mobbing behaviour as the stressed parents become increasingly desperate in their attempts to drive the dog away. Broad, white wings flash like angry semaphore as each bird takes turns to dive. The dog's owner arrives on the scene, but to my absolute astonishment she encourages it to chase the birds, talking to it in the loud voice that most people reserve for very young children: 'Yes, clever boy, yes, you love to play with the birds, don't you? Clever, clever boy!'

Matt and I look on in open-mouthed disbelief as she rewards the dog's behaviour with a treat from her pocket, before striding on to spread the joy elsewhere. We walk on in horrified silence, each locked in our own thoughts as the birds alight on the short grass next to the track. One slinks straight back towards the hidden nest, while the other, presumably the male, stands guard in the open off to one side.

The breeding population of lapwing has decreased by 50 per cent since the mid-1990s. Obviously this is not entirely due to unruly dogs, with habitat loss and intensive farming being key drivers, but incidents such as this certainly don't help. My childhood seems to have been filled with enormous winter flocks of these wonderful birds in the Avon valley but, now I come to think about it, the only decent flock I've seen recently was swooping over six lanes of motorway as I drove north past Stafford one grey winter's day. Like many ground-nesting birds, the peewit seems locked in a swift downward spiral that too few of us even seem to know about.

Friday 8 May

A dry start. The forest leaves hang with mist, but the sky is clear and the 4 a.m. journey to goshawks is sublime. The full moon hangs huge and heavy above dark spires of conifer – a Spielberg backdrop in need of a flying silhouette. The road through Whitemoor Glade is a bright bridge of silver and I turn off my lights to follow, rolling slowly forward as the trees eventually rise up to swallow the moon. As I re-enter the darkness, a tawny owl is perched on a sign next to the road. It ignores me and remains poised with head tilted forward, listening to something in the leaf litter. I switch off the engine in the hope of watching it hunt, but it seems to notice me for the first time and flies off into the shadows.

Further up the road I emerge back into the open beneath the looming silhouettes of two giant redwoods. Memories of climbing them when I was younger come swooping back. This is one of the highest ridges in the whole forest, so I can only imagine what the moonlit view from those high tops must be like on such a beautiful night as this. The forest seems utterly deserted. The open heath beyond is dark but sits beneath a lightening sky and by the time I reach goshawk wood the stars are fading, and birds are starting to call.

The dimly lit wood is awash with song as Matt and I walk quietly in. Once again, I keep my movements slow and gentle as I ascend. The goshawks must be getting used to the routine by now, but I can't afford to get complacent. After such an exposed climb, entering the darkness of the hide comes as a relief. Kneeling to catch my breath, I tune in to the sounds of the wood. A flushed gos will usually call from a distance as it flies to and fro looking for an opportunity to return. All is peaceful, so I risk a slow lift of the flap of my hide's window. A shadow in the heart of the nest tells me the female is still there.

As we haul up the camera and Matt then walks slowly away, the dawn chorus is in full flow. The barely perceptible sway of the tree, combined with being so high above ground, makes it feel as if I'm floating through a stream of song. Blackbirds, thrushes, blackcaps, robins and wood warblers all pour out their melodies. The cutty joins in from his tree-root pulpit and I'm especially happy to hear the distant cry of the curlew again. Cadences swirl and eddy through the trees while two cuckoos throw their simple refrain back and forth from both ends of the wood.

After a time, I hear the distant honking of geese approaching from the west: a call of the wilderness drifting down through centuries. The geese are still a mile or so away, but their honks echo across the heath and up through the woods to where I'm crouching. They're flying low, far lower than usual, and I wonder whether they know there's now little to fear from the light aircraft normally buzzing through the skies above Bournemouth and Southampton. I close my eyes and listen to the approaching sound of air being beaten beneath strong, tireless wings. Time stretches, elongating, but finally they fly directly above me, the woods reverberating. Even the other birds seem to pause to listen, and I sit spellbound as they slowly fade into the east.

It's been well over ten minutes since I first heard them. Ten minutes of the purest, most evocative sound imaginable. The insidious clamour of the twenty-first century feels like a world away and not for the last time I wish it would stay there, forever banished to an alternative reality.

I spend my life trying to get away from the clutter of modern life. I don't think I'm alone in finding its relentless demands trying. I'm not talking about the people around me – quite the opposite in fact. What I struggle with most is the constant barrage of noise, advertising and digital stimulation that has become accepted as the norm. It seems deliberately designed to corrode and undermine our ability to recognise what truly matters. We live in strangely disingenuous times and I often wish that fewer distractions lay between me and what counts. The trick, I guess, is knowing what really matters in the first place. Family and friends, obviously, but a personal relationship with the natural world also plays a huge role and it's no coincidence that, like many, I have always turned to the woods for help when needing to reconnect or recalibrate. As the novelist John Fowles put it, 'Slinking into trees was always slinking into heaven' – my bridge to sanity and a broader mental horizon. So, the experience of being in the forest during such extraordinary times as these is nothing short of magical for me. But it is also surreal – like being visited by the Ghost of Forest Past, who reveals evocative glimpses of what once was.

Before the Normans arrived, this region was known as Ytene; 'the place of the Jutes', referencing the Germanic tribe who arrived in the wake of the Roman departure. It must have been an exciting, wild place full of mysterious links to a deeper Celtic, Neolithic or even Mesolithic past. The wail of the curlew and trumpeting of geese would have been commonplace and I wonder at how we've arrived at a point where something that was so intrinsic to a landscape could become so rare and provoke such bittersweet emotions. The sound of a flock of geese in flight taps into deep genetic memory and somehow stitches the history of a place together. But it is also tinged with regret, a powerful reminder that we have polluted the airwaves just as much as we have the earth. We've grown so accustomed to filtering out background noise that we barely notice it any more. Reconciling what we see with what we hear is often impossible. We've flooded the sky with so many radio waves and carved it up with so many machines that a world without our acoustic clutter no longer exists.

And so it is that listening to that skein of geese traversing an empty sky has become one of my most precious moments in this strangest of springs.

The male goshawk calls his mate off the nest for food and I press record. I can almost feel the warmth spill from her fluffed-up breast as she stands, stretching stiffly before dropping off into the wood. It looks as though it's been a chilly night and a moment later the male bounces in to take her place while the bed's still warm. Being so much smaller than her, he has difficulty spreading his legs wide enough to straddle the large eggs. The nest seems to swallow him whole and if I hadn't seen him arrive I'd never know he was there at all. The female takes her time eating whatever the male brought and I wish I knew where the delivery spot was. I'd love to get a shot of the birds away from the nest in a different setting, but therein lies the rub. They're not called 'Phantom of the Forest' for nothing.

Trying to film a goshawk away from its nest can be a thankless task. They tolerate a hide near their home only because they are already committed to the site and have little choice but to get over their reservations if they wish to breed. The nest provides an intimate and unique window on their world, but for a short time only, and when such an important focus is removed from the equation free-ranging predators such as the goshawk can be extremely unpredictable. Yet although it is a limited perspective, it is the only one I've got.

Half an hour later the female returns with a full crop and a beak full of dead pine needles. Mysteriously, she drops them onto the back of the male lying beneath her. Research suggests that raptors often bring sprigs of greenery back to the nest to help control parasite loading. The green foliage of many conifers contains anti-bacterial compounds that actively disinfect and help keep the nest clean. I've filmed many forest raptors doing that, but never seen a bird return with dried, dead needles like this before. They don't add any structural integrity to the nest, nor, being so old and stale, would they have any antiseptic benefit. Is she simply adding to the nest lining (if so, why not wait until the male has left), or even trying to camouflage the eggs? It's strange behaviour that for now remains a mystery.

Having sprinkled her mate with needles, she remains standing and starts to call vociferously. The male is still lying low beneath her, pretending she's not there, but she's clearly anxious to get back on her eggs and trying to displace him without having to resort to giving him a kick. Her tail coverts puff up and I notice how the downy white feathers sometimes reflect the mood of these otherwise enigmatic creatures. She takes another step forward, raising her huge powerful feet to disentangle her talons from the twigs she's standing on. She's calling non-stop now: *kek-kek-kek-kek-kek-kek-kek*, throwing great effort into making as much noise as she can. Then, when she's practically standing on his back, the male hoicks himself up and slopes off reluctantly, as if he's jealous of his mate's closer bond with their clutch. The female steps carefully into the deep cup, bending down once more to turn the eggs one by one with her beak. When she eventually sits, it is with infinite care and immense satisfaction. Her muscular back rolls from side to side as she settles lower and lower and I'm reminded of a chicken-shaped egg basket we used to have in our kitchen when I was a kid.

A few hours later, the male calls her off the nest again, but this time he's barely had time to get comfortable before she's back. Once again, she has a beak full of dead pine needles, but this time she leans forward and presents them to him. He ignores the offering but gets up to leave. The female drops the needles as before, letting them fall idly from her beak. She doesn't bother arranging them in any way. It's all a bit casual and strange, and the only explanation I can think of is that she's using them as props to help her shoo the male away.

I don't see him again today. The female stays on the nest all afternoon, getting up only to stretch her left wing out and down, opening her broad tail feathers like a fan to reveal their dark stripes and subtle contouring. This goshawk camouflage is a masterclass in disruptive patterning, playing on the relationship between sunlight and shadow and helping to break up the outline of their body. Counter-shading also helps the bird blend with the sky when viewed from below and the ground when viewed from above.

Predators such as goshawks don't just have to keep concealed from their target prey, they also need to keep a low profile to avoid detection by any other creature tempted to raise the alarm. I've seen plenty of big cats forced to give up their hunt, not because they've been spotted by their quarry, but because their cover has been blown by other animals who kick up such a storm that it's only a matter of time until the prey itself gets rattled. In fact, some birds have specific alarm calls for specific predators, as anyone who's watched a flock of swallows mob a hobby will know. One of the best ways to find a cheetah in the long grass of the savannah is to listen out for helmeted guinea fowl. These birds create such a disorientating commotion that I've seen cheetah almost beside themselves with anger and irritation. Once the game's up, the predator – be it hawk, cat or anything else

– generally either moves on or simply starts preening or grooming itself, as if to indicate haughty indifference.

I also suspect that the barred goshawk plumage helps confuse their prey. Goshawks can really shift when they want to, and an ambush is usually launched at blistering speed when the animal is momentarily distracted. Motion dazzle is created when bold patterns move rapidly, making it difficult for predator or prey to judge speed and direction accurately – a phenomenon employed by zebras to help evade lions for example, although big cats have reduced colour vision, whereas birds have heightened sensitivity to colour.

The female continues her al-fresco yoga: wing stretches known as mantling followed by a quick round of warbling, the excellent term for the arching of folded wings above their backs. Being nest-bound for thirty-five days must be quite the ordeal.

Saturday 9 May

We're back on the Dartfords, but in order to give the previous birds some space, Matt and I decide to try a different part of the heath. Moving further along the ridge through a patch of recently burnt gorse, we take up position near the silent railway line and are treated to another blinding sunrise. The receding shadows reveal burial mounds against a red sky, and I am reminded that nearby Latchmoor literally translates as 'moor of corpses', perhaps in reference to the Bronze Age tumuli or in memory of an ancient battle reputedly fought here in the sixth century. Despite the portents, the morning's filming goes well, although the warblers here don't seem to have an active nest and are more preoccupied with chasing off rival intruders.

We switch location for the afternoon, heading north to where I saw my one and only Dartford as a kid. It's a long shot, but it pays off. There's clearly still a population here some thirty-five years later and we get a few useful shots of the birds foraging and singing alongside linnets and stonechat. We even catch a fleeting glimpse of a red-backed shrike – a big tick for me. Once seasonally common, the 'butcher bird' stopped breeding in the UK thirty years ago for reasons that remain unclear, although a pair or two have successfully bred in Devon within the past ten years. I'd give a lot to film one at its larder, a miniature gibbet stocked with captured prey impaled on thorns: usually a hawthorn or blackthorn bush, decorated with the grisly remains of lizards, insects, mice and small birds, like some sort of ghoulish Christmas tree.

Monday 11 May

A blustery start as I sit outside my own home with a cup of tea. The wind has switched to a northerly and is arriving in frigid gusts that I track through the valley by the bending of distant trees. A flurry of willow leaves spins over from a neighbour's garden and our chickens retreat to their coop as if they know what's coming. I decide to stick it out, watching the tall, slender silver birch I planted eight years ago flex like a gymnast and relinquish its catkins to the wind like confetti. I hate to see blossom and flowers wasted this way, but only a fool gets angry with the weather, so I turn back to my rapidly cooling drink and wait for my phone to load last night's official update from Downing Street.

The video buffers while Boris Johnson's frozen face looks haggard and tired. I wouldn't want his job. A poisoned chalice if ever there was one; it can't be easy. Even so, by the end of the speech, I'm left wondering what on earth he was trying to say and am amazed he made it through without keeling over from an acute case of gesticulation. As I drain the ice-cold dregs from the bottom of my cup and am finally driven back indoors by the worsening weather, I also wonder what effect such a confused and ambiguous message is going to have on our country.

I'm out for a walk with my eldest when my phone rings and 'Andy NF' flashes up on the screen. Rohan and I are in the middle of the local woods and it's blowing a hooley. Ash trees thrash wildly around above us, spilling their leaves to the wind, while the oaks lean into the gusts. We've come up to check on the local buzzard nest, but I'm worried about how things are down in the forest. Preparing myself for bad news I answer the call: 'That goshawk nest in the larch blew down last night.'

For a brief moment it seems my worst fears have come to pass, but then I realise Andy's talking about the first location we visited in mid-April – the huge nest teetering in the top of a skinny larch close to a road. It had been my fallback option if our first choice hadn't come through, so I'm relieved we hadn't had to resort to it. Thinking that there must be another nest in that territory, I ask if the birds will lay again.

'Don't think so. That's it for them this year, I should think. Three chicks. Two survived, but the third was killed.'

This brings me up short. I'd assumed that the fallen nest contained eggs like our filming site. Then I remember how late in the season our chosen female started to sit. Other nests had a head start on ours, but even so I'm surprised when Andy tells me the two surviving chicks are already showing pin feathers. This suggests they're over two weeks old, maybe eighteen days, which puts the destroyed nest at least three weeks ahead of ours.

'Where are they now?'

'Over at Martin's. He says you're welcome to go and film them if you like.'

Wednesday 13 May

I'm keen to get a few shots of the downed nest. So, while Matt goes to make sure our own survived the storm, I load the van and head off in the opposite direction, having travelled back down here last night. The storm has long since passed, but the roads are still covered with debris: spindrift leaves, broken branches, and I pass several downed trees, including a huge oak that has crushed the railings on an old bridge. Fresh piles of woodchip show where limbs have been removed to open up the road again.

The sun is up and slanting through the trees by the time I arrive at the fallen nest, barely five metres inside the woodland. These birds never cease to surprise. It isn't often that you get the chance to look at a gos nest so closely, but it's a poignant scene. The gale clearly ripped through the trees here and the moss-covered floor is strewn with wreckage from above. Snapped branches lie everywhere, two trees are down and the nest itself sits in a heap at the base of the spindly larch that once held it. It's huge, well over a metre in diameter and almost the same in depth. Kneeling down among the scattered lichen-encrusted twigs, I roll it upright. It stays intact, bound tightly by the clever way it's been woven, like wicker. It's a lump of a thing and I'm amazed that any of the chicks survived the fall. The two that did must have been thrown clear; the third crushed.

The deep cup that so recently held the focus of so much care and attention is empty save for a few scraps of wispy white down. More clings to the nodules on twigs around the edge and a large adult feather is lodged deeper. I tease it out and smooth the ragged gaps in the barbs before holding it up against the rising sun. Rotating the quill slowly between forefinger and thumb I watch the light play through the thin vane and marvel at its perfection. A distant alarm call from an adult bird reminds me of their trauma and I feel slightly ghoulish in my fascination for their fallen nest, but can't help delving deeper to see if it also contains any remains of their prey. It yields two sparkling-blue jay feathers and the fleshy foot of a wood pigeon. Further down, the twigs from yesteryear are soft and decomposing. Tiny flea-like insects squirm and ping to escape the light and I find a bony mouse jaw and skeletal squirrel's paw that seems to greet me with a macabre high-five. These dark depths are clearly home to a fascinating ecosystem all of their own. I rearrange the twigs and fluff up the nest as if to redress the damage from its fall.

Nests in trees other than larch rarely grow to these monstrous sizes. The rows of nodules so characteristic of larch twigs are formed from old buds, and so when woven together these gathered pieces lock to provide a solid, interconnected framework on which the birds build each season. The whole thing grows organically, year on year, until it comes to resemble an eagle's eyrie more than a hawk's nest. Such nests can be used for a decade or more. I once filmed one in the Forest of Dean that was 1.5 metres across and 3 metres deep, a veritable treetop fortress that eventually subsided in bad weather. I've never seen one blown clean out of a tree before though. I can only assume that the one at my feet was undermined by weak supports and sure enough, looking up into the larch, I can see the snags of thin broken branches glowing orange in the morning light.

Setting the camera up on a set of sliding rails, I get a few tracking shots of the nest as the sunlight comes and goes between the crowded tree stems behind. As I'm packing up, I notice a fresh splash of white nearby. At least one of the adults was here just before I arrived, probably perched on a branch above to peer down at the fallen nest in confusion. Although this pair can build another, or even commandeer a buzzard's, it still represents the end of an era for them. I wish I could at least let them know that two of their precious chicks have been saved and stand a good chance of survival.

In fact, I'd go as far as to say that those two chicks are now in the best hands possible. Martin and Julia Noble live nearby and over the thirty-four years they've lived in the forest, their secluded house has become an oasis for animal rehabilitation and a focus of energy for conservation. As an ex-head keeper of the New Forest himself, Martin has long been a proponent of balanced woodland management and actively helped put wildlife firmly on the agenda during his thirty years with the Forestry Commission. I've known Martin on and off for a long time, having pestered him for advice as a teenager, and I can't think of two better people than him and Julia to take on the challenge of raising two rapaciously hungry goshawk chicks.

Opportunities to get close to wild goshawk chicks don't come round very often. Realising that the nest will provide a useful backdrop, I grab a spare blanket from the van, lay it flat on the ground and roll the nest on top. Taking hold of the four corners, I wrap up the bundle and lift it onto my back. The nest is of no further use to the birds, so I don't feel too guilty as I steal out of the wood with it hanging over my shoulder like a burglar's swag bag.

Half an hour later Martin and I are standing an antisocial distance away from each other as we chat in the fresh air of the paddock next to his house. I've just caught sight of an orphaned badger in the fenced enclosure behind us, and can hear a jay screeching, presumably at the sight of two pine martens in a pen in the back garden. A wren flits down into the ivy at the base of an old work shed, its full beak telling me there's a nest in there, and a great spotted woodpecker chips from an oak above. The whole place is alive and even through my face mask I can smell the pungent aroma of a fox's recent visit. The property has the comfortable, rustic feel of a nineteenth-century frontier homestead. Ingrained within its environment, it's the kind of homely dwelling where trees lean in to see what's happening and the lines between man and nature are blurred. In fact, it's difficult to know where the property ends and the forest begins.

Carrying the nest through the stable yard, I place it on top of a crate balanced on a wheelbarrow and trundle the whole thing into the spring sunshine. Julia brings out the two chicks, placing them gently in the fir-lined cup of their old home. They're in good health despite their recent ordeal. Clear-eyed and alert, they immediately peer up into the sky directly above to check all's clear. Pretty standard raptor paranoia, even from an early age it seems. Julia is obviously doing an amazing job with these two. In fact, they're so full of beans, it takes her a moment to get them settled as they take turns to teeter on the edge of the nest and look quizzically at the wheelbarrow's handles. When the larger of the two looks ready to jump, Julia covers them both gently with a small towel to settle them down. They duly comply and a few minutes later the towel is removed, and I get my first proper look.

One chick is slightly larger than the other. This could be the difference between female and male or could simply mean that it was the first to hatch. In study sites, the ratio of males to females is established via ringing the chicks. Although sex can sometimes be determined from fifteen days old, most ringing is done from three weeks. Chicks, or 'pulli' to give them their proper name, are occasionally ringed on the nest itself, but usually lowered to the ground where detailed measurements of defining characteristics, such as body weight and leg, beak and feather length, are recorded. One of the most striking differences between female and male goshawks, even at this age, is the pronounced thickness of the female's leg. Adult females are significantly bigger than males but, even so, the increase in leg thickness and foot size seems out of proportion. Female gos have muckle great feet with significantly longer talons – all the better for killing squirrels and rabbits, prey too large for most males to tackle.

Both chicks are covered in grey fluffy down, speckled with the newly emerging tips of their first proper feathers. Their backs and wings are fringed with these small rectangular plates, which look a little like roof shingles. Each one is chocolate brown, tipped with chestnut, while the emerging tail feathers look like tiny puffs of orange flame on the end of grey pin feather sheaths.

Focusing my macro lens on the face of the larger chick gives me an opportunity to study the close-up detail that I could never hope to film from a distance at our wild site. The most striking feature is its eyes. Not orange or red like an adult's, but icy grey, as if frozen by the northern wind that blew them out of their nest. Their brows are already deeply ridged and tiny specks of grey fleck the side of its face. A bright yellow cere, the waxy patch housing the nostrils at the base of the beak, brings an incongruous splash of colour, and the long, yellow mouth stretches right back to end in a strangely sardonic upward kink, like a sneer. The beak itself is dark grey and rippled with growth rings like the shell of a fossilised devil's toenail. Despite its vulnerable age, the overall impression is one of supreme confidence and purpose tinged with precocious aggression.

My eye is drawn to its bright yellow feet. Reptilian and prehistoric; scale-plated in horny ridges that expand and contract as the bird flexes talons to grip the twigs on the nest. The three forward-facing toes end in needle-sharp claws about a centimetre long, but it's the fourth, rear toe that really draws the eye. At least twice as big, it's perfectly designed to push through prey like a blade. Falcons either kill their prey on impact, or with a powerful bite to the neck. Goshawks, however, grapple, stamp and squeeze, wringing the life out of their prey until it stops struggling. No neat snip to the neck here. Sometimes they don't even wait for the animal to die before beginning to eat.

By the time I've finished filming, Andy has arrived and the four of us stand around the two chicks discussing their future. Andy suggests placing them with surrogate wild families and already has a suitable nest in mind for the smaller of the two chicks. Both birds have full crops, so we decide to at least get the first one settled while the weather is on our side.

Leaving the larger sibling with Martin and Julia, Andy bags up the other, places it on his passenger seat and we head off in convoy across the forest. I've never done this before – placed an orphaned goshawk chick in with a foster family – but they are adaptable birds and Andy's confident it will work.

It's not until we're almost at the tree that I realise I've been here before, during our epic search a month ago. There are plenty of larger trees around, but for some reason this pair of goshawks has chosen a gangly old Scots pine. It's tall but skinny, and the nest is built in an awkward nook where the main trunk divides, 50 feet up. It's woven around both stems of the fork and there are no branches of any size above it to support a rope. I don't like using climbing spikes on Scots pines as their skin is thin and easily damaged, but it feels like the only real option if I'm to avoid damaging the nest or the chicks in the process.

An adult goshawk calls in alarm from somewhere close by. Pulling on my harness, I strap on my spurs, flip a lanyard round the tree trunk and step up onto my spikes to get the job done as quickly as possible.

Climbing up to a raptor nest containing chicks is very different from visiting an empty one at the start of the season. The adult birds now have something precious to defend, have no idea as to your intentions and will sometimes risk an attack, coming in fast to strike the back of your head. Several years ago I was knocked unconscious and nearly killed by a harpy eagle that sank its talons down into my neck next to my carotid as I fitted a camera on its nest in Venezuela. It was entirely my own fault for underestimating the bond that had developed between mother and chick. It was a valuable lesson, and even though that female harpy was many times the weight of a goshawk, I wouldn't want to place these New Forest hawks under such stress that they also felt the need to resort to such desperate measures.

I've spoken to people who have been attacked by wild goshawks while ringing chicks at the nest in North America, but have not heard of anyone experiencing such a robust reception here in the UK, which is strange. Although the American birds are a different subspecies, they look almost identical, inhabit the same niche and originally descended from the nominate Eurasian species, so we might expect the same kind of treatment here. I can only assume that this difference in behaviour stems from the fact that our own goshawks descend from feral populations, and that such innate aggression towards humans had been diluted down through the generations as part of selected breeding by falconers. Still, it doesn't pay to get complacent. With one eye on what's going on behind me, I begin to climb.

The bark might be thinner than a larch's, but the timber's a lot denser. The spurs don't penetrate nearly as far and I find that as long as I move with deliberate care, avoiding half-hearted stabs, I can avoid any real damage. Halfway up I lean back to take a quick rest while peering up at the nest above. I can't see any of the resident chicks, but then wouldn't really expect to. They're around a fortnight old but lying low. We just have to hope that our orphan is of similar size despite having been fed extra well over the past few days in captivity.

Continuing on my way I arrive at the base of the nest but it's just too big and bulky for me to leapfrog. So, shuffling up as high as I can on my spikes, I climb above my safety lanyard and wrap my arm around the skinny right-hand fork. I'm now eye-level with the top of the nest and able to peer over its rim for my first look at its occupants. Three chicks are arranged top to tail in a deep bed of pine needles. They look healthy, though are hunkering down to avoid catching the attention of the strange hulking creature that's just appeared in their midst. The whole tree is shaking from my additional weight, so I quickly pull up the bag containing our orphan. Opening it just enough to get my hand inside, I invert it and the chick tumbles softly into my waiting hand, the bag falling to the ground far below. As I turn the chick upright, its long legs dangle down between my fingers and I carefully lower its huge feet onto the nest. Threading my abseil rope through the fork, I transfer off my climbing spikes and prepare to head down. My last sight of the orphan is of it sitting on its haunches surrounded by its new siblings. We've given the chick the best chance it could hope for.

Thursday 14 May

Andy's offered to show me a couple of curlew nests. It's early morning and I'm making my way across the forest to meet him. Seven weeks after lockdown began, yesterday saw the start of official easing. So far it's making little difference to the low number of people out and about and it's still deserted, but I can't help feeling that the pendulum is about to swing the other way.

Curlews are the UK's largest breeding wader – a striking bird that's easily overlooked due to the simple fact that it's always been part of our landscape. Its soul-searching voice is the sound of summer on the heath and winter on the seashore, but it's in trouble.

There are currently around 66,000 breeding pairs in Britain, with the majority found on the upland moors of Scotland and northern England. This may sound like a lot, until you realise that this represents 25 per cent of the entire global population and that the sixteen-year period from 1994 to 2010 saw a decline of 46 per cent in UK breeding pairs alone. The entire population is contracting at unprecedented and unsustainable rates. A 17 per cent reduction in breeding range in Britain is bad enough, but the shocking 78 per cent decline in breeding numbers seen in Ireland is nothing short of a disaster. So it's no surprise that in 2015 the curlew was added to the UK red list of endangered birds. Populations appear to be collapsing simultaneously across the board, suggesting multiple issues are to blame, making things tricky from a conservation point of view. When you consider that, in the RSPB's words, the UK is 'arguably the most important country for curlews in the world', the whole scenario is extremely worrying and not a little depressing.

With its close proximity to coastal wintering grounds, the New Forest has traditionally been a good place for them. Yet, over the past twenty-five years the numbers here have plummeted in line with everywhere else. With only around 300 pairs of these traditionally upland birds now nesting in the whole of the British lowlands south of Birmingham, every nest counts. It's a story that seems to epitomise current worldwide trends in biodiversity and from an ecological standpoint – as well as a moral and emotional one – it's devastating. Few birds provide such a direct link between coast and countryside – an important cultural, biological and spiritual theme that I would argue is enough to warrant their protection. Also given the UK's crucial role as a focal point for the global population, if we can't continue to provide safe refuge here, then where else can they go?

I meet Andy in the north of the Forest, not far from the goshawk nest, in a flowering hawthorn copse. We inevitably fall to chatting about how the 2020 breeding season is going and whether the lockdown has so far brought any recognisable benefits to the birds.

It seems too early to know, but as Andy leans against the door of his 4×4 and pours himself a cup of coffee from a Thermos, he gives me a better sense of how things stand in general terms.

There were just over a hundred pairs here when Andy first began working in the area, thirty years ago. They seem to have stabilised a bit at around forty-five pairs, but if you look at the simultaneous declines elsewhere, it's obvious that there are a lot of things going on with curlew that aren't good. The climate crisis is likely affecting these waders quite considerably, but on the wild moorland places, such as the New Forest, Exmoor and Dartmoor, there's also considerable predation of the chicks from crows, ravens, foxes and badgers.

Andy explains that because wader numbers are lower in the forest than before, the birds aren't able to drive away some of the predators collectively as they once did: 'If you have a mire system with twenty-five pairs of lapwings on it, you'd get fifty lapwings chasing a crow out of the area. Collectively they could help protect that colony, but nowadays there's only one or two pairs on these systems. Same for the curlew.'

And so once the chicks hatch, unable to fly – a period of vulnerability that lasts a long time for these large birds – the mortality rate is high, and that's where Andy thinks the tipping point comes in.

There's a pause in the conversation while a blackcap joins in to give us his opinion from the hidden depths of the brambles on the other side of the track and I ask Andy what can be done at a local level here in the forest.

'Those are some of the things we're trying to tease out at the moment,' he says. 'We're going to have to work so hard to keep them here, and I just don't know how sustainable that is without some sort of predator control and without the will of the public to curb their behaviours during the nesting period.'

I ask him whether lockdown has given the curlews any respite, for this year at least. Like this whole picture, it's complicated. Many bird species leave the forest for most of the wintertime, returning to take up their territories only in the late winter and early spring, but how many birds are excluded from potential nesting sites because of visitors is hard to quantify.

'A bird comes back, it looks for a suitable place; it starts to sing, holding that area. It assesses the level of predators, the level of disturbance and what it thinks are its chances of rearing young at that site. If it doesn't feel right, it moves on. These birds, including the curlews, would have selected their territories before lockdown even kicked in.'

Global warming, high local disturbance and high numbers of predators. The odds seem stacked against the curlew and while the birds might be able to cope with any one of these in isolation, the combination of all three is obviously proving increasingly insurmountable.

We lock our vehicles and leave the blackcap to continue the discussion on his own as we skirt through the willows on the bank of a stream and cut out across the heath. Stepping off the sandy trail, we make towards a low, pine-covered hill. The nest is down in the bottom on the other side.

The heather is knee-deep and as we approach the brow of the ridge Andy tells me they'll be off the eggs as soon as we go over the top. We won't have long.

Sure enough, as soon as we break the skyline my eye is drawn to a small, hunched shape scuttling away through the marsh grass on the edge of the mire below us. It was only luck that I saw the bird at all and in a moment it's melted away, leaving no visible clue as to where the nest might be. As we emerge from the pines onto the edge of the marsh, Andy begins quartering the faded yellow tussocks, eyes to the ground. I follow in his wake, not wanting to step on unsearched ground for risk of accidentally damaging the eggs. In truth, having never seen a curlew nest up close before, I have no idea of what I'm looking for beyond the obvious. So I keep behind Andy and tread very carefully.

A few moments later he stops abruptly and, sidestepping to the left, points down into the long grass. There, in a shallow bowl of pale woven grass, lie three of the most beautifully marked and perfectly camouflaged eggs I've ever seen. They certainly knock the goshawks' pale, chicken-like offerings into touch. These are a rich olive with dark chestnut mottling like some sort of exotic granite. Beautifully cryptic and subtle. In fact, if it hadn't been for the long strips of blond grass on which they lay, I suspect even Andy would have struggled to find them. The other thing that surprises me is their size. They're massive; the size of turkey eggs. There are three, but Andy's confident they'll lay a fourth before incubation starts in earnest. This means they'll hatch a month from now, if spared, so we leave the nest in peace before our presence draws the interest of the crows that spend a lot of time perched on trees in the heath, watching for any opportunity. This is one of Andy's research nests, part of a comprehensive survey of curlew in the forest this year. Having confirmed that the eggs are intact and viable, we retrace our steps to leave them in peace. Our fleeting visit has caused no harm and, having been left undisturbed to lay their eggs during lockdown, this particular pair now stands a reasonable chance of raising their young. I just hope they manage to remain below the radar as the country begins to open up again.

Returning to our vehicles, I fall into line behind Andy as we drive south to the second curlew site he wants to show me. Halfway across the forest he slows down in the middle of the deserted road. I pull alongside.

'Want to see a woodlark nest?' he asks, elbow leaning on his open window.

From a neighbouring lay-by, we step onto a closely cropped sward of grass that flows like a golfing green between clumps of gorse and blackthorn. Like the curlew, woodlarks are a ground-nesting bird, but much smaller. Once again, I move behind Andy to avoid making an unforgivable faux pas. I'm beginning to realise just how ignorant I am about certain British birds. I get to know certain species extremely well during filming, but often stay embarrassingly clueless about others. Ask me a question about the dietary prefer-ences of lowland gorillas in the Congo and you'll get a tediously lengthy answer, but ask me to describe the difference between an icterine and marsh warbler and there'll be a lot of sucking of teeth.

This time, however, even Andy seems stumped by the nest's camouflage. Muttering quietly, he circles back on himself several times until he eventually finds what we're looking for. Gently lifting up a skirt of heather, he beckons me over to take a look. Nestled in a nondescript bowl of dried grass lie four unbelievably delicate, vulnerable-looking chicks. My shadow falls over them and, assuming that this heralds the arrival of food, they raise their heads to beg. The curlew nest seemed vulnerable enough, but the location of this one in the short grass right next to the road strikes me as almost suicidal. It's hard to imagine a more exposed setting and while it is well camouflaged to our senses, a dog would find it in a New York minute. It's also hard to believe that it could survive with so much livestock lumbering around the open forest. The whole nest would fit neatly beneath a pony's single hoof, but as I say this out loud, Andy explains that livestock actually help create and maintain the environment these birds favour.

'They're quite short-legged – the birds that is, not the cows,' he chuckles, 'and like the short, heavily grazed areas. If the vegetation gets too long then the woodlarks would disappear. Their success is linked to the level of grazing in the forest and because there are high levels at present, it's suiting the woodlark.' And as for being trodden on? 'It's funny,' he says. 'You'd think they would get crushed. But you'd be surprised how careful a cow is about where it puts its feet. They seem to sense the nest is there and avoid it.'

It's nice to hear of a positive relationship between wildlife and livestock management, but I can't help thinking that the real threat here, once again, is disturbance from people – and often their dogs. This year the lockdown must have helped, but what about in more usual times? Andy tells me that while areas like these are often busy, woodlarks are pretty determined little birds: 'Because the adults sit very, very tight on the nest, they get away with it quite a lot.'

All well and good, but I can't help thinking that for some dogs I've known, an adult bird sitting tight on its eggs would simply be added protein. I realise that the majority of dog walkers are sensitive enough to put their animals on a lead when it counts, but for many (including my younger self, I hasten to add), a walk in the forest is an opportunity to let their dog run free, often out of sight. And for most of the year, why not? One of the problems I guess is that the nests of ground-nesting birds are so discreet, it's a case of out of sight, out of mind. If people aren't aware of them in the first place then how can they be expected to do anything about it come spring? Most of us would be mortified at the thought of our dog or bike wheels destroying a nest or initiating its predation once an adult bird's been flushed. So, raising public awareness is a large part of what Andy and his colleagues at Forestry England do and, sure enough, there's a bright red Forestry England sign nailed to the wooded barrier at our next location: 'Stop – rare birds nesting HERE now! . . . Please keep only to main tracks. Dogs should be on the tracks too – if necessary use a lead. Disturbance to the birds will leave eggs and chicks vulnerable to predators. Thank you.'

Leaving our vehicles outside the gate, we follow a track across the heath, past a wide, low pond surrounded by rock-hard mud pocked by the hooves of ponies. Flanking a thicket of flowering gorse, we emerge on the side of a valley. The ground falls away in front of us to reveal a far-reaching patchwork of heath and marsh below. These overlooked but expansive margins are some of the forest's most precious and unique landscapes, truly prehistoric and extremely susceptible to damage from development and modern intensive farming. Islands of Scots pine punctuate the scene and drifts of white bog cotton show where an extensive mire lies hidden beneath the dun-coloured grass.

There are only 120 valley mires left in the whole of Western Europe and no fewer than ninety of these are found right here in the New Forest. It seems all the more ironic that such a marginalised bird as the curlew should be driven to seek refuge in such a threatened landscape and not for the first time I marvel at how the forest has managed to survive to the present day with so many of its precious habitats intact. Perhaps we have something to thank the Conqueror for after all.

A familiar warbling lifts our eyes towards the vast empty sky as a curlew glides down. Tracing its effortless descent, I watch it circle wide and low over the nest site. I'm encouraged to see that the closest track through the heather is at least 300 metres away from the site and that there is a convenient gorse-covered knoll where I can set up my hide without drawing attention. Close enough for my lens, but far enough for their comfort.

Heading back across the heath, we pass a curlew feeding in the shallows on the opposite side of the pond. With a nod in its direction, Andy tells me he recognises it as the male bird from the nest we've just visited. I ask him how he can be so sure and am told to watch. A few seconds later the bird tries to step forward but seems to falter and stumble before regaining his balance. A look through the binoculars shows he has a broken right leg. He can rest lightly on it, but his foot is curled into a ball and the knee thickened and immobilised by scar tissue. It doesn't seem able to bend, so every staggered hop involves an exaggerated swing of the limb that's clearly very awkward. 'He was on this territory last year,' says Andy and I wonder at how such an injured bird could have survived a British winter or even recovered from such a nasty injury in the first place.

Driving back to digs, I realise with a sickening lurch this might be my first and last chance to film these wonderful birds in the forest.

Of the forty or so established territories in the forest this year, most – if not all – will finish the season without a young bird to bolster their numbers. The presence of a territory doesn't automatically mean chicks, or that they'll be allowed to fledge. Corvids, foxes, dogs, bikes, people . . . we're all playing our part. And while it's easy to blame predators, it also occurs to me that crows don't have a choice about how they behave. We do. So, in this instance it seems sensible for us to accept responsibility and be mindful of our actions. Surely these beautiful birds are worth fighting for and if this means keeping pets on leads or not riding our wheels over heathland in spring then it seems a small price to pay. Because if things go on as they are, the New Forest curlew may well become yet another dreadful absence in this landscape within the next twenty years.

Friday 15 May

The days are getting longer. Sunrise is now 5.15 a.m., and it's light enough to see from four. As long as we're at the base of the tree by 4.15, with a bracing alarm clock at 2.45, I can be set up and good to film from five. The first few minutes after sunrise are critical. Long, solid shafts of light push through the dark canopy to bathe the nest in gold.

The first rays appear at 5.18. Almost horizontal, they hit the larch 10 feet above the nest and immediately start seeping down the trunk. Only now do I notice that the nest is empty. The female had been there, I'm sure of it, but must have sloped off in the pre-dawn chill without me noticing. A few minutes later, as the rays hit the nest, I hear a goshawk calling from my right. A shadow plumes vertically up through the trees and the female materialises back on the nest. The woven twigs cushion her long legs with a bounce, and she leans forward to inspect her eggs, before turning to face east, towards the rising sun. The pale feathers of her breast catch fire and her eyes flash gold. Worth getting out of bed for.

The rays move on and dissolve in the growing light of day. The female lies down to doze. Turning off the camera, I sit back in the darkness of my hide and listen to the forest waking up around me. A flock of long-tailed tits flits unseen through the branches above. One of them hovers to pluck an insect from the side of the hide, its wings a blur as its silhouette comes into focus on the backlit canvas.

The day grows warm, the wood grows quiet and the goshawk sleeps. Head forward, eyes closed; beak resting on twigs.

Trying to stay awake myself in the stuffiness of my enclosure, I feel my phone vibrate against my chest. It's Matt D, the forest keeper. I can't pick up, but listen back to the voicemail: 'Good morning, James. Sorry I haven't been back to you – they put me on furlough for a fortnight, so I've only just come back to work, but I've got a litter of fox cubs for you. There's half a dozen of them running around, not far from where we were before. So, give me a call if you're still interested.'

I immediately text and word comes back that he's around this afternoon.

It's obvious that the goshawk chicks haven't yet hatched, but it's also obvious from the female's behaviour that it won't be long. Twice I've filmed the female rouse herself long enough to peer down inquisitively at the clutch, standing with chest fluffed out as she cranes her head down to stare and listen. Based on our estimated lay date of 11 April, we're now thirty-five days in, so can expect the first egg to hatch very soon. The fully formed chicks may even be chirping from inside their eggs by now, indicating that they have broken through into the tiny chambers of air at the top of their eggs in preparation for 'pipping' the egg and pecking their way out fully a day or so later.

I make the difficult decision to call off filming the goshawks to go and scope out the foxes. I already have plenty of footage of the eggs being incubated, and even if the chicks did start hatching today, they'd lie limp, low and exhausted. It's a big nest, with a deep cup, so it will be several days before the tiny balls of fluff are strong enough to reveal themselves.

I start packing up carefully in preparation to leave without disturbing the female, her eyes half closed in the warm dappled sunshine. I haven't done this before. I've always arrived under the cover of dawn and left in the gloaming of dusk, so I'm about to break the rules. Goshawks don't like that. They don't like surprises.

It's long been obvious that the adults can hear my movements within the hide. It doesn't worry them too much, but as the day wears on and the sun swings round behind me, I'm silhouetted and they can see my shadow. So, if I'm going to move, I need to start now. It's warm today, so the eggs wouldn't cool too quickly if I scare her away temporarily, but I also need to be aware of other predators that might seize the opportunity to raid the nest. Unlikely, but not impossible.

The biggest threat is from buzzards. They've been active over the wood, riding the thermals above the trees as the days have grown warmer, and must be nesting somewhere nearby. Their wailing calls drift down through the canopy and the goshawks peer up nervously to track them across the sky. No raptor enjoys being below another. It makes them feel uneasy and vulnerable. Large shadows occasionally wash across the nest and although a buzzard would be no match for a goshawk, especially a female as large as this one, it would make short work of the eggs or newly hatched chicks given half a chance.

As I'm packing, my wedding ring taps the metal of the camera tripod and this alone is enough to elicit a fierce glare. Her eyes lock onto the hide and I sit still until she relaxes and settles back down. It takes me an hour to get squared away and at 11.30 I text Matt to ask him to walk in. A few minutes later her head flicks to the left and a jolt seems to pass through her body. She's tense and fully alert. I watch as she tracks Matt's movement through the trees towards us. A few moments later I hear the familiar crunch of dry needles and twigs and, peering down through the slit at the back of the hide, I see him 50 feet below, eyes upwards. Turning back to the female, I give her a few minutes to get accustomed to this unusual midday intrusion. She's not happy, but neither is she freaking out. I wouldn't have dared try this a few weeks ago, but putting faith in her ability to recognise our exit routine, I begin lowering the heavy camera.

I let the rough rope slide through my hands inch by inch. I lower it slowly, avoiding sudden movements while trying to stop the bag twisting erratically. The pulley line is rigged on the back of the tree and the tall, straight trunk conceals most of the movement, but it's a big bag and I am worried about it startling her as it spins side-on. It takes five minutes to lower it into Matt's waiting arms. My rucksack and tripod follow without any issue, but the sight of me emerging is just too much for her. I leave via the back flap, so that the hide and trunk conceal me pretty well, but she sees my feet dangling in space and a low whistle from Matt tells me she's flown. Since the game is up, I get down as fast as possible. I shoulder the camera bag; Matt grabs the rucksack and tripod, and we walk quickly, quietly away. I'm aware of being watched.

Halfway back to the car a faint *kek-kek-kek* tells us she is back on the nest. I look at my watch. She's been off for seven minutes. On a hot day like this, the eggs can cope with up to forty-five minutes of exposure, especially since the chicks inside are so large and close to hatching. Matt tells me she tracked each load down to the ground, then stood up stiffly to throw us a cursory glance before leaping off. I am happy with that. It was unlikely she'd stay put the whole time, so for her to leave in such a calm manner is a result.

There are a few cars now parked up in lay-bys, but these look like local people seeking out favourite haunts for a stroll now they can travel further from home. The wider forest is still deserted as I squeeze my van past ponies dozing in the middle of the empty roads to meet Matt D in a still-empty gravel car park. The springtime sun is shining, and the sheer abundance of life around us is staggering. The world is awash with beauty. Every bush has its own nesting song thrush, every bramble patch its resident blackcap and male greenfinches sing from the blossom-covered branches of every crab apple. Or so it seems.

Matt D and I are discussing the foxes and I'm reassured by his patient approach.

'The cubs haven't been above ground long and still have that wonderful smoky-brown coat. Like wolf pups.'

It's a great image and I can't wait to see them. From what Matt tells me, it sounds as if they're around six weeks old. Fox cubs spend the first four weeks being suckled by their mother in the underground chamber or tunnel in which they were born, their wide, soft tongues forming a perfect seal around her teats as they lie nuzzling with eyes closed in the comforting earthy darkness. During this time the vixen will barely leave their side, relying on the dog fox to deliver food in much the same way as a male goshawk provisions his mate as she broods. The vixen will venture above ground again at around the same time as her cubs open their eyes. When she's not there, they lie in a pyramid of tiny chocolate-brown bodies to conserve heat, but a week or so later they too begin to emerge above ground, their slate-blue eyes blinking in the harsh sunlight of an exciting world completely unknown to them.

A squabbling sound makes us both look up into the old oak above us. Starlings are nesting in a hole and four tiny beaks are poking out in jostling competition for their next meal. Above the starlings, blue tits probe the newly opened buds and leaves of the upper canopy in their never-ending search for caterpillars. There's nothing quite as insatiable as a brood of young blue tits, except perhaps a brood of young great tits. And, sure enough, both species are foraging side by side; beaks delicately collecting tiny plump green bodies in a futile attempt to placate their offspring. I've never understood how birds manage to open their beaks to pick up something without dropping everything else they're already holding. Consider puffins, for example. It's one of life's wonderful mysteries, I guess.

A familiar red car pulls up alongside me and we all head down into the woods in convoy.

It's good to be back in the north of the forest, an area I love. It's steeped in history, both natural and human. We pass the place where I discovered scraps of Roman pottery when I was a boy, thrown up on to a spoil heap by badgers. One of the pieces – a dull grey handle – had the clearly defined smudge of a thumb where it had been moulded to the side of a jug. A 1,600-year-old thumb print is enough to make you think hard about your own place in the universe.

Soon the three of us are walking deeper into the woods as quietly as the tinder-dry leaf litter will allow. Without raising his hand to point, Matt nods slowly off to our left beyond the shadow of a large silver fir. Young bracken shoots poke up through a jumble of fallen dead branches and there, staring straight at us, is a six-week-old cub the size of a kitten. There are several more playing in the bright dappled sunlight behind. With eyes locked on ours, it steps silently forward, its pointed ears and square face blending perfectly with the bracken. True to Matt's description, its coat is a smoky grey-brown, the very image of a wolf pup. Its fur is woolly with long guard hairs backlit by the afternoon sun slanting through the oaks high above. Magical. We back off to leave them in peace before they smell us and bolt underground.

I give it thirty minutes before heading back in on my own, this time flanking to the left to allow for wind direction. Scent is everything, whether they can see you or not: one whiff of human and they're gone so fast it's as if they've been swallowed by the ground. Crumbling a dead leaf into dust, I let the powder drift on the air to make sure I've got it right. There's a gentle prevailing south-westerly, but this will eddy in such dense woodland so regular checks are needed. Stepping off the track between two magnificent haw-thorns in full bloom, I use an old enclosure ditch for cover as I crawl through the leaves on my belly in search of a good vantage point. I am now separated from the fox earth by a stream in the bottom of a narrow, steep-sided trench. Crawling further forward to find a spot with a reasonable view, I lie on my front while slowly building a makeshift hide out of dead branches. Having draped camouflage scrim over the low framework, I crawl back out to the van to collect the camera kit. Once all is in place, I let things settle before raising my binoculars.

Scanning the woodland for the cubs, I see them heaped on top of each other in a patch of sun off to the left. I'm in the wrong position. Squirming out, I slowly work my way further along, using tree buttresses as cover. A hundred feet later, I'm rewarded with a clear view of the sunbathing infants. They're out for the count and I watch the rise and fall of their tiny bellies, the smoky-quartz-brown of their fur blending perfectly with the thick blanket of fallen winter leaves. The warm sunshine reveals a hint of ginger in their coat too, but the cubs won't start to take on the familiar red of their parents for a while yet. Apart from the occasional flick of a tail or sprawling stretch, you'd never know they were there; a scene of perfect contentment and peace. These cubs were born in early April at the height of lockdown and judging from the relaxed curiosity with which we'd been greeted earlier on, I'd say we're some of the first humans they've seen. They haven't yet learned how dangerous our kind can be.

It's a wonderful view, but I'm too exposed with no adjacent deadwood or trees for cover. Even though they're sleeping, I have no doubt that their senses are tuned into the sounds and smells of the wood. I don't want to give the game away by dragging deadwood through the leaf litter, so crawl on towards a couple of young firs I can see growing on the banks of a ditch, standing at the confluence of two small brooks. Their branches mesh together just above the ground to provide a discreet screen through which to film.

It takes another hour of painstaking crawling and moving of kit to relocate fully, but by 4 p.m. I'm set up in the shadows beneath the trees with a clear view of the earth a hundred feet away. Not as close as I'd like, but although the cubs seem oblivious for now, I have no doubt that their mother is lying somewhere close by. She might be only a few metres away, curled up between the moss-covered buttress roots of an oak, or she may be napping in the bracken on the other side of the ridge. Her chosen spot might change from day to day, but either way she'll be keeping a close eye on them.

The cubs may be naive, but vixens are anything but, and the mother of these cubs will immediately move them to a different den should she grow suspicious. She won't want me anywhere near them, so I can't push it any more than I already have. To be honest, I'm amazed I've made it this far.

I've set up the tripod in the ditch below me so that the camera is above ground, but underneath the low-hanging branches. A natural pillbox. I can sit safely out of sight in the bottom of the trench with the rest of the kit beside me, slowly standing up to scan the woodland in front with binoculars. The wind continues to eddy through the trees, so the ditch also helps contain my scent and I spend the next few hours crouching below ground level, trying to avoid getting soaked by the water flowing through the mud beneath me. Though I can't feel my legs any more, there are compensations. As the evening light softens and the day begins to cool, the pile of legs, tails, ears and noses begins to untangle as each cub wanders off in search of something to do. This gives me an opportunity to look at them properly.

At this age, their coats are woolly and fleece-like; their tails tapered like a calligraphy brush. A buff point already hints at the distinctive white tip seen in some adults. Their legs are darker than their flanks, and their throat and chest are a shade lighter still. Their muzzle is shorter than an adult's – all the better for suckling – and the backs of their pricked ears are dusky black. Unsurprisingly, telephoto camera lenses give a false sense of size and while I'm able to fill frame with a single cub, I also know that they are small enough to fit into my cupped hand. I don't like the word cute, but if ever there was the epitome of such a thing, then a fox cub is it. You can keep your domestic kittens.

Cute they may be, but tame they aren't, and – like the goshawk chicks – the most arresting thing about them is their eyes. In time they will harden into flint-edged amber, but for now they are clouded by a wash of blue-grey that reflects the gaps in the oaks high above. They appear distant and foggy, an effect heightened by the way the cubs sit staring into space as if trying to focus on something only they can see.

I keep filming as the cubs jump and roll around the mossy roots of a huge oak. Like so many infants, these youngsters love playing king of the castle and I realise I'm smiling as I film them pounce, tumble and scrabble for the high ground. Another thing I notice is that all play, however boisterous, is conducted in absolute silence. Not a peep. They are so light-footed they make no sound at all as they grapple in the dry leaves, rolling over bluebells and knocking bracken.

They are so endearing that as I follow them through the lens, I wonder why young mammals elicit such an emotional response from us. No doubt a lot of this is to do with vulnerability, but even the cubs of feared rivals (and I'm speaking on an evolutionary level here), such as bears and big cats, still elicit this desire to connect and protect. I suspect that the domestication of wolves – the early ancestors of today's dogs – was born of this paradox. Looking down the camera, I wonder whether this could be due to something as simple as the dome of a cub's forehead, which closely resembles the profile of a human baby's – an effect accentuated by forward-facing eyes and small button noses. As altruistic and heart-warming as this is, there is also a solipsistic slant to it. We place our own physiology on a pedestal and are naturally drawn to animals that remind us of ourselves. We are a narcissistic and capricious bunch when it comes down to it. And since – for better or worse – we are also the planet's dominant species, plenty of other animals we purport to love and respect suffer at our hands as a result. The illegal pet trade in protected wildlife stands testament to this.

Our cultural relationship with foxes is a complicated one. Cubs soon become unwitting receptacles for all manner of historical and cultural human baggage as they grow older. Until one day, as adults, they find themselves adored by some and loathed by others. Although not legally classed as vermin, they are often treated as such and I can't think of any other large British mammal that is not afforded at least some protection from being shot, if only seasonally. Foxes have none, even during the breeding season. Which, in my mind, makes them all the more special. They are the architects of their own success and owe us absolutely nothing. Foxes seem to exist in spite of us. Or at least many rural ones do. They are true survivors and I'm grateful for this window on their secret world as I film this roiling tumble of cubs at play.

Saturday 16 May

The oak wood is cool and dim when Matt and I arrive. Keeping chat to a minimum, we are preparing to creep in with the equipment when I spot a familiar shape lying on the track ahead. I must have dropped my cap on my way out last night, but placing it on my head I'm almost knocked out by the stink of fox piss. Taking it off, I see the dark patch on the material and stupidly take a closer sniff to confirm the obvious. The acrid reek of rotting ammonia fills my nose, making me gag. The dog fox must have come across it during the night and decided to eradicate the offending stench of human in the best way he knew how. I'll give it a wash tonight, but for now it may even help disguise my own scent, so I pop it back on my head to Matt's disgust.

Stepping silently between the two blossoming hawthorns, we follow a deer trail into the wood. The two conical firs look like witches' hats from a distance. Placing their dark silhouettes between us and the fox earth beyond, Matt and I crawl towards the ditch beneath them. Having slid quietly down the bank into the ankle-deep water, I set about readying the camera. I plan to stay until after dark this evening, so Matt wishes me luck and leaves me to it.

It's now 6.45 a.m. and the wood is alive with birdsong. The main chorus is soon over, but the blackbirds, wrens and jays continue to call, now in alarm; their clucks, rattles and screeches gathering momentum until the whole wood is filled with their hysteria. Such a racket usually indicates the presence of a predator, so they've probably seen one of the adult foxes nearby.

I'm wrong. It's a tawny owl. The jays flush it out of its ivy-clad day roost, and it flies off into a neighbouring oak. Looking through my binoculars, I can see it perched on a snag staring back with fathomless black eyes. Its plumage is a collage of hand-painted feathers flecked with every shade of brown. It's soon spotted by the jays again and this time I lose sight of its broad, round wings against the shadow of the deeper wood.

Tawnies must get tired of being constantly mobbed, but I can't say I blame other birds for wanting to move them on. They know perfectly well what tawnies are capable of. It's not just mice that fall prey to them in the dead of night. I've seen footage of tawnies bringing back jackdaws, magpies, and many other creatures to feed their chicks. Roosting birds make easy targets and tawnies pack a punch. They're also silent and pretty much invisible with it. In fact, from a jay's point of view, tawny owls can be a total liability. The ultimate neighbour from hell. So, when one's discovered hiding up during the day, it's firmly escorted off the premises before the tables turn back in its favour at sundown. Tawnies can even be aggressive to people who venture too close to their nest. The pioneering wildlife photographer Eric Hosking learned this the hard way when he lost his left eye to one in 1937. In the depths of night, the screech of a tawny must be enough to freeze the marrow in tiny rodent bones.

While I've got the binoculars to hand, I scan the fox earth, paying close attention to any small piles of leaves that may turn out to be sleeping cubs. It's still early and cool, so I assume they're safely underground dozing and digesting food brought to them during the night. Sitting down on the stool I've brought in, I'm still hidden safely below ground and, although the stream runs over my boots, I can stretch my legs out fully and stare up into the branches. Definitely an improvement on the stifling claustrophobia of the goshawk hide.

I'm halfway through a cup of tea when a deep, drowsy drone announces the arrival of a queen bumblebee. Truly enormous, she is laden with treasure sacks of pollen. Swaying gently from side to side, she hovers in front of a hole in the bank of the ditch. I worry I might have damaged her hidden nest while setting up my tripod but am happy to see her land on a patch of moss then crawl purposefully into a dark crevice, her tiny black legs moving mechanically like a clockwork toy. Leaning forward for a closer look, I see a small round hole leading directly into the bank of dry earth. It's an old mouse or vole burrow, and I imagine the solitary queen crawling slowly down the tunnel to nurse her handful of precious eggs in the musty darkness of the royal chamber beyond. These endearing insects have almost birdlike nesting behaviour. It's still early in the season, so she'll be working hard to raise her first brood. Unlike honeybees, bumblebee workers don't overwinter and so it takes a queen quite a time to build a new colony each spring. The golden pollen strapped to her hind legs will form a tiny pedestal on which she lays her first eggs. Next to this she makes a small amphora-like honey pot out of wax. This she'll fill with nectar to sip regally while brooding her eggs

– wrapping her large furry body around them to keep them warm. A week or so later the larvae will hatch, feed on the pollen pedestal for a few days, then pupate to emerge eventually as fully developed adult workers around thirty-five days after the eggs were laid. It's a complex process that takes longer to complete than it does most birds to incubate and raise their brood of chicks.

The dry clatter of wood-pigeon wings pulls me out of my sub-terranean reverie. Two of them are trying to balance on the same branch of ivy in an oak directly above the fox earth. It's too early in the year for there to be any berries, so I can only assume they're nesting there. It's the same patch of ivy vacated by the tawny, so I wonder what dramas have gone on at their nest during the night.

As the morning wears on, the woodland settles and sunlight hardens from gold to silver. The air around me hums with insects and I put on a mosquito head net to thwart the relentless midges. Worryingly, I can't smell the fox's offering on my cap any more. It's amazing what you get used to.

The fox cubs appear just after 11 a.m. There's no sound, just a feeling that they are now there. I peek above the lip of the trench to see a tiny bundle of bodies and turn on the camera. The open oak woodland forms a natural arena around the den and although the unfolding bracken will soon occlude it, the gently rising ground is visible for now. I count five cubs. A group of three grapples together in the leaves to the right, while the other two lie sunbathing at the entrance to the den in which they were all born. The three at play are in shade, but as I watch through the camera one of them rolls into a patch of bright sunlight, sits up and stares straight at me with its iced-over eyes. As it turns its head away, the sunlight melts through to reveal warm amber irises below. A glimpse of what they will become.

As I pan across to the other two cubs, one sits up and I can see that it has a misshapen nose. There's a large lump on the bridge; it looks broken. It also appears to be the runt of the litter, although from the way it's pestering its sibling to wake up and play, it hasn't suffered much as a result. It's full of beans; gently but relentlessly chewing its sibling's muzzle until it has no choice but to defend itself with a half-raised paw, which is also immediately bitten and tugged. This seems to do the trick and the dozing cub is pulled fully awake. The game is on and rising on tiny hind legs, Broken Nose body-slams its sibling, which rolls onto its back and kicks up with all four feet. Gentle teeth grip each other's throats as they roll over and over before tumbling down the spoil heap at the den's entrance. The other three take interest and soon the five cubs are chasing each other through the flowers, crouching, pouncing, ambushing and tackling in exactly the same way the cheetah cubs had done in Kenya, two months and a pandemic ago.

Eventually tiring, they leave the game in turn until the last is left sitting on its haunches wondering where they all went. The others are now in full-blown exploration mode, fanning out in different directions to muzzle through leaf litter, gnaw deadwood and paw inquisitively at unseen distractions.

As if I'm trying to herd cats, I struggle to keep track of which cub is where. One of them disappears down a hole, only to emerge from an entirely different one a few moments later. Foxes don't really dig breeding earths with multiple entrances and exits, so the vixen is probably raising her cubs in an old badger sett. There's an active sett close by, so it makes sense. Badger setts can be huge, rambling affairs, growing over generations as tunnels and chambers are gradually added or blocked off. Some of the oldest and biggest can be well over a century old, if not more.

Gerald Lascelles was deputy surveyor of the New Forest from 1880 to 1914, and in his book *Thirty-Five Years in the New Forest*, he has this to say about how local badger setts are often structured: 'In the forest many of the earths are made in a stratum of sandy soil, beneath which is clay or boggy and wet grounds. So that the earths do not run more than 8 or 9 feet deep, but often spread over as much as three-quarters of an acre, with innumerable entries, galleries and passages, all communicating with one another over this extent of ground.'

Quite the underground mansion. An endearing image embellished further by the following: 'Often there are two storeys of such galleries, one running above the other, and the badger moves from his ground-floor apartments to his first floor as he thinks . . . best.'

A dyed-in-the-tweed sportsman, Lascelles seems to have hunted most things during his tenure in the forest. Such were the times. As deputy surveyor he also brought a degree of much-needed order to the unruly hunting scene. Although deer, foxes and otters continued to be hunted with hounds, albeit under new regulations, he seems to have had no time for badger baiting: 'I cannot see where the sport comes in . . . and I would never sanction any such proceedings, nor allow a live badger to be taken away from the forest, to be used for a purpose of that nature.'

Just as brutal on the dogs as it is on the badger, baiting had already been outlawed for forty-five years by the time Lascelles arrived in the forest, but I have no doubt it still continued (as it continues elsewhere to this very day, sad to say). Lascelles clearly saw it for the cowardly activity it is and seems to have had a bit of a soft spot for old Brock: 'When I first came, I found there were a good many and for thirty years I never allowed them to be destroyed, deeming them, generally speaking, harmless creatures, such as ought to be protected in a State Forest amongst its other denizens.'

Is it too fanciful to see the seeds of wildlife conservation and enlightened forest management in this last sentiment?

Lascelles' 'Gentleman in Grey' is generally a clean, hygienic creature with rarely any mud on him, despite being such a relentless digger. By comparison foxes can be rather untidy, smelly neighbours. Badgers won't hesitate to block off tunnels and annex foxes should they move in during the breeding season and from the way the fox cubs seem to be popping up from three or four different holes in the ground, it seems that this particular family is squatting in a disused wing.

Just to add to the Kenneth Grahame-esque charm of the situation, the ridge on which this entire oak wood stands is riddled with the remains of an ancient Roman settlement. Romano-British potteries thrived here 1,600 years ago and although formal excavations tell of humble wattle huts and rustic kilns, these woods were once alive with the shouts of people, the clink of discarded wasters and the pungent smell of woodsmoke. Like most geological maps, that of the New Forest looks like a dog's breakfast to me, but it's clear that this particular region held ample supplies of everything a Roman potter could want: sand, clay, fresh water and plenty of timber to fire the kilns.

The shards of broken pottery I found as a boy came from here and it's hard not to think of enduring old Badger in *The Wind in the Willows* as he explains to Mole how his underground home is excavated from the ruined buildings of an ancient people, long since vanished: 'People come – they stay for a while, they flourish, they build – and they go. It is their way. But we remain. There were badgers here, I've been told, long before that same city ever came to be.

And now there are badgers here again. We are an enduring lot, and we may move out for a time, but we wait, and are patient, and back we come. And so it will ever be.'

The New Forest potteries were modest affairs, so my daydream of badgers lying on beds of silver and shuffling past buried marble statues is probably wide of the mark. Still, the fact remains: these people passed through, but the forest endured. As a metaphor for rewilding it's a strong one that resonates all the more during these strange days of lockdown Britain. People come. People go. But nature remains. I'm left wishing that it really was that simple.

By mid-afternoon it's hot out there beyond my ditch. The sun has swung round to slant beneath the branches above me. Shafts of gold illuminate fine particles of dust suspended in the air and, since the cubs have disappeared again, I pass time by watching the swirling patterns that form when I blow gently. A whirr of wings announces the arrival of a coal tit on a dead twig a few feet away. It either hasn't seen me or simply doesn't care I'm there. I keep still and watch as it cocks its head from side to side, spins round on its perch, then drops down to bathe in the brook 6 feet away. Again and again it bends forward to splash its head and nape before fluttering its tiny wings to make sure all of its feathers get cleaned. Looking as though it's just been through a spin cycle, it flits back up to its perch where it starts to preen meticulously. This is the start of a steady stream of tiny woodland birds coming to bathe. Blue tit, willow tit, chaffinch, blackbird, goldcrest and – most exciting of all – a crossbill; all take turns using the same spot while totally oblivious to my presence. I can't believe I haven't been rumbled and I realise that it's not only us humans who sometimes see only what we wish to.

As afternoon rolls on into evening, the sunlight becomes gloopy, flowing over the leaf litter like syrup. Then, right on the very edge of dusk, when all colour has drained from the wood and I barely have enough light to film by, the vixen arrives to check on her cubs. She lopes down through the trees from the north to be greeted by a chorus of excited whickering and tiny twirling tails as all five run around her feet and jump up to lick her muzzle. She lies down to suckle but is up and off again within a few minutes. Just before she disappears at the top of the slope, she turns to look straight at me. There's no concern in her expression, but she's letting me know that *she* knows there's something here in the shadows. She couldn't have seen, heard or smelled me, but there's more to a fox than the five basic senses common to us all.

Sunday 17 May

It's still dark in my hide and I'm writing these words in the faint tiger-stripe light filtering through its canvas. Mornings are noticeably lighter now, but it always feels like twilight here in goshawk wood. The journey in was beautiful. The sky on fire. Red and purple mirrored in silver-still heathland pools. The tawny was calling from the same oak at the wood's entrance and I caught another glimpse of the resident roe buck sloping off between the enclosure's fence wires. It hasn't rained for two weeks and the van lurched and wobbled as it came to a halt on the sun-baked ruts next to the now familiar tree stumps and fallen logs. Broken branches jutted up from the hard ground like bleached bones in the half-light and the grass verge was dotted with tiny star-shaped yellow flowers that weren't there two days ago.

By 4.55 a.m. Matt and I were standing at the base of the platform tree; the female was sitting notably higher in the nest, much more upright and prominent than before. As I inched my way carefully up the 50 feet of rope, I hoped that this meant the chicks were finally here.

It's just after 6.30 a.m. and the male stands motionless. Blood-red eyes gleaming in the dawn light as he peers down at something hidden beneath him. All the signs are pointing to one thing and I picture the newly hatched chicks in my mind's eye. Three or four puffs of white swaddled by fresh sprigs of fir. The male remains frozen as he stares down with his impenetrable gaze. A few minutes later the female returns and he leaves, leaping off into the labyrinth of branches beyond. She stands too, as if trying to reprogramme her brain to come to terms with these strange new arrivals. Eventually emerging from her trance, she spends a few moments preening her breast before lowering herself in stages; moving carefully from foot to foot as she tries not to smother or squash the tiny scraps of life huddled beneath her.

Watching this finely tuned avian killing machine display such cautious maternal instincts is extremely endearing, but I can't help thinking how she must look to her cowering offspring. Like staring up at a looming T-Rex.

In truth, the chicks are still too young to be thinking about much at all. I imagine that their thoughts simply mirror their most pressing immediate needs: a stream of consciousness rotating gently between sensations of hunger and temperature. At this stage they have barely any idea of what or where they are. Their sleepy days filled with a gentle flow of primal urges.

A short while later the male materialises on the edge of the nest for the third time this morning. This time he's brought food directly to the female. A blackbird fledgling, although it's hard to be certain. It's almost impossible to identify most of the corpses that arrive on a goshawk nest, especially since the male usually plucks and butchers them beforehand. It's like trying to recognise an animal from the inside out. In these final stages it is simply meat. A package of life-giving protein, still warm, with a stomach full of worms and insects fed to it by its own parents only minutes before. I know from experience that now the gos chicks have hatched, things will really ramp up on the hunting front as their father brings in a steady conveyor belt of small bodies, leaving the larger mammals such as rabbits and squirrels for his powerful mate to tackle in due course. Soon she'll also be able to slip the bonds of the nest to indulge her pent-up desire to hunt for the first time in weeks.

The male gos has spent countless hours watching the patterns and movements of the wood's feathered residents, like an expert poker player biding his time as he studies his opponents before going in for the take. He'll know every single songbird on his patch. In all likelihood he will have deliberately avoided killing some birds until they have had a chance to raise a brood of their own, at which point he will casually drop by to steal one chick after another, followed by their parents too if he can.

It is this kind of goshawk behaviour that many people – gamekeepers and conservationists alike – sometimes struggle to come to terms with. A few years ago, one pair of New Forest goshawks even cleared out every chick from a hobby nest. Nevertheless, for every 'special' bird that gets predated, there are dozens, if not hundreds, of other so-called pests that are also consumed. Crows, jays, wood pigeons, rooks, rats, rabbits and all the rest. To say that a goshawk's diet is eclectic would be an understatement and although no one wants to see a hobby nest raided, such occurrences are incredibly rare and pale into insignificance next to the huge number of grey squirrels, for example, that are taken.

Love them or loathe them, grey squirrels do a lot of damage. The stripping of bark for food, nesting material and as a source of trace minerals can kill a tree. Hardwoods such as beech between the ages of ten and forty years are targeted most frequently and once a tree has been ring-barked it will die. Red squirrels can be just as bad, but we don't have them in the New Forest any more, their former range having been taken over by the larger introduced greys. It also seems that greys can live in higher densities than their smaller cousins and this in turn causes additional gnawing damage via the aggression-displacement behaviour often used by males to avoid direct conflict. The environmental impact caused by high local densities of any single creature (including humans) can be really significant.

Unsurprisingly, the numbers of New Forest squirrels, red or grey, were traditionally controlled for the sake of valuable timber resources. In the nineteenth century, New Forest oak contributed to Britain's 'wooden walls' and one of Nelson's own commands, the sixty-four-gun man-of-war *Agamemnon* was built here on the banks of the Beaulieu river. That New Forest timber played such a prominent role in the Battle of Trafalgar is an amazing thought, whatever your patriotic leaning. Today, foresters use shotguns or traps, but two hundred years ago squirrel control was a more prosaic affair involving gangs of men lobbing weighted cudgels called 'squoyles' and 'snogs' at the hapless rodents. In 1889 alone, some 2,200 red squirrels were culled in the forest, not that many of these went to waste. Squirrel pie was once a popular local dish, particularly at Christmas when the squirrels are generally at their most active.

Goshawks, then, are a natural means of squirrel control. A female goshawk with a brood of hungry chicks can go through the local population without mercy. In this, the goshawk is the forester's friend.

It's at this time of the year, when the pressure is on to hunt continuously to feed their chicks, that the goshawks' supreme design and character comes into its own. The wide, rounded wings that catapult them through the smallest of gaps in foliage; the long, stiff tail that fans out into a broad rudder to help the bird jink and brake with blistering speed and also provides extra lift when the wings are tucked in to squeeze through a gap. And of course, the long legs and powerful feet that keep the thrashing prey well clear of the hawk's precious eyes and primaries. In my opinion, it is these incredibly long legs that help set the true woodland hawks aside from other raptors. Like its smaller cousin the sparrowhawk, the goshawk sometimes uses its legs to reach in to probe a bush or burrow for prey gone to ground and I once watched a grounded sparrowhawk stalk round a garden shrub before reaching in to pluck out a terror-frozen chaffinch.

There is also evidence to suggest that many species of raptor deliberately avoid hunting in the vicinity of the nest until they absolutely have to. The equivalent of keeping a larder stocked for a rainy day, I guess. Or maybe even a way of ensuring their chicks don't have to run the gauntlet of poaching in another goshawk's territory when learning to hunt for themselves. Either way, the male goshawk now has four or five mouths to feed and it is his time to shine.

And shine he does. Within an hour he's back again, this time with a plucked, dressed and headless wood pigeon, which the female stands to receive before lowering her beak to the carcass. Her movements are slow, deliberate, even tender as she tears tiny strips from it to feed her hidden chicks. As she leans forward, I catch a glimpse of two tiny gaping beaks through the nest's woven ramparts. Their mother is careful to feed each in turn, turning her enormous beak sideways as if it were a pair of forceps to place each morsel delicately directly into their waiting mouths. The tiny scraps of pink meat

110

disappear as quickly as she can pluck them and the to and fro continues for at least twenty minutes. Eventually, the tiny jostling shapes sink back down out of sight and their mother leaps away with the pigeon's pitiful remains dangling from one set of talons. There was still plenty of meat on the carcass, but she returns empty-clawed a few minutes later. I presume she's stashed the remains somewhere, or simply left them for her mate to finish. It makes total sense not to keep meat in the nest itself, of course, as it wouldn't take long to putrefy and the last thing the chicks need is an infestation of blowflies.

On her return, she again prepares to brood the tiny plump bodies beneath her – an endearing display of gentle clumsiness: tentative movements as if she's afraid to touch the chicks at all, yet those muckle great feet and talons of hers are not exactly designed for delicacy. Nor is her grappling-hook beak for that matter. It can't be easy to avoid impaling your chicks while carrying a foot full of carving knives. It's captivating to watch, but nerve-racking none the less.

By midday the nest is flooded with bright sunshine and the back wall of my filming hide glows warm to the touch. The new mother can hardly keep her eyes open and, giving in to the inevitable, her head slowly drops forward onto her chest. I struggle to avoid doing the same in my stuffy, airless chamber.

Monday 18 May

It's early morning and I'm heading back to Somerset for a few days. The traffic is already very heavy, very fast, and I soon pass a freshly killed fallow doe lying on the hard shoulder. Her beautiful spotted flank is swollen and even from a distance it's obvious she's heavily pregnant. She hasn't been there long, and I slow down to make sure she isn't simply stunned, but as I pass I see a magpie feeding on her demolished skull and exposed brain. She was only a couple of weeks away from giving birth and was clearly hit hard, probably as she ran across the road in a blind panic. Her unborn infant is still inside her. There's nothing I can do, but can't shake the harrowing thought of her fully grown fawn slowly dying in the womb as it listens to cars rushing past barely inches away. It makes me feel sick.

I pull over to call Andy to let him know it's there. He sounds stoic. 'Why can't people slow the fuck down?' I vent. 'I don't know', comes the simple, resigned answer.

What I was forgetting of course was that he's had thirty years of this. Roadkill is a fact of life here. He's seen it all hundreds of times before and long since come to terms with it. I don't think that much surprises a New Forest keeper. Their role often brings them into contact with the best and worst of human behaviour. Pragmatism is not only required for the sake of their job but is also essential if they are to stay sane. I realise there's a lesson to be learned here. There is no moral high ground. That doe could have just as easily run out in front of my van and scenarios like that are often in the lap of the gods. I dare say that Andy was simply relieved he didn't have to deal with a three-car pile-up. Better a pregnant deer than a family of four.

Nevertheless, I can't stop thinking about that poor fawn. Nurtured in its mother's womb for eight months only to die a sordid lingering death at the hands of a driver halfway through their morning latte. Seems so unnecessary, so cavalier.

The recent message from Downing Street regarding the easing of lockdown has had a huge effect on the forest. No one seems to know what the rules are any more and the amount of traffic on the roads since the 15th has been shocking. Worse than before, if that's possible. A large portion of the country is still off work or on furlough but can now leave home and the forest has been inundated by a tsunami of vehicles.

It was truly incredible to witness how quickly nature reclaimed space from us during the early days of lockdown, but now that the pace of life is ramping up again, it feels all too dreamlike.

Thursday 21 May

I've been lying low, enjoying a couple of days at home. The clouds of white blossom have now been replaced by lush green foliage and as I open the gates to leave home once more, I'm greeted by the earthy smell of red valerian, also known as fox's brush. It's a strange plant, smelling like honey to some and cat's urine to others. Unfortunately, I'm in the latter category and those growing below our hedge always seem most pungent around sunrise.

The roads are quiet and misty at first but entering Bath I find myself sitting in my first jam for months. People have been allowed to travel for work and exercise for the past week and the roads are now busy, the traffic fast and aggressive. A lot of pent-up energy and frustration is being released: people overtaking on blind corners; angry cyclists and suicidal motorbikes. By 7.30 a.m. I'm part of a depressing train of contractor vans and quarry trucks winding its way south towards Salisbury. Stopping in my usual place to get petrol, I put on a mask and gloves before entering the shop. I am one of the few wearing PPE and receive irritated looks. Two blokes in their late twenties stand so close behind me that I ask them to step back and remind them they should be wearing masks. They remind me to mind my 'own fucking business'.

Arriving in the forest, there are no spaces left in the gravel car park, but I spot Matt's red car parked up by the forestry gate. I pull alongside and he tells me that his half-hour drive has taken him an hour. The forest hasn't been this busy for months and it's barely 9 a.m. on a Thursday. A knot of cyclists huddles around a map on the corner; several groups stand around rugs laid out on the grass; dog walkers stride past, frowning in pursed-lipped disapproval. Neighbours greet each other with smiles as their dogs wag tails, but their faces shut down again as they pick their way past the strangers. Once the ice is broken, we English are generally a friendly bunch, but we can be pretty stand-offish until then, especially when it comes to our 'territory' – an island-born idiosyncrasy that isn't helped by the fear and suspicion sown by a global pandemic.

Unlocking the gate, Matt and I leave all this behind – feeling as fortunate as ever – and head down the long track into the company of trees. As we park up on a grassy ride beneath a yew, birdsong floods in and I step out into the sacred glow of a springtime oak wood.

The bluebells are going to seed; their heavy heads sag as they lean drunkenly against each other. The warm air is now filled with the spicy camphor of unfurling bracken. Each stem seems to stretch and yawn as if waking up from hibernation and their tightly twisted fiddleheads are covered in soft silver hairs that shimmer like silk.

I breathe deeply and feel tension and anxiety ebb away. I hadn't realised how wound up I'd become. I've grown accustomed to having all this space to myself and this morning's journey had coiled me up as tight as a spring. A quick cup of tea does wonders to unravel my own fiddlehead and by the time I've unloaded the camera kit I'm more or less in the right space to concentrate on filming.

I'm no sooner settled in my ditch when a newly fledged black-bird slips down the dusty bank into the water behind me. Its mum calls in alarm from somewhere nearby, so crouching low to avoid breaking cover, I shuffle up the ditch and use cupped hands to gently lift the bundle of soggy fluff out of the brook. Its yellow mouth gapes up at me as it squawks and wriggles. Reaching over the top of the bank, I place it safely in the leaf litter beyond, then retreat to watch from the comfort of my stool. The chick shakes itself and sits down, all but invisible in the leaf litter. A few minutes later it starts flopping towards a clump of last year's bracken and I see its mother swoop down and bounce along behind with a beak full of wriggling worms.

The distant echo of a cuckoo draws attention to the morning's stillness. This isn't the uncanny, watchful stillness of a goshawk wood, more the companiable energy of a million living things peacefully making the best of the spring sunshine. With no sign yet of the cubs, I do my best to melt into my surroundings.

I've always enjoyed filming from hides. I love the process of silently watching and waiting for things to happen in their own time. Arthur Cadman, another New Forest deputy surveyor turned author, summed it up perfectly: 'The rewards for those who can train themselves to sit still and hold themselves quiet are great.'

Nature doesn't respond well to being pushed but can be incredibly generous when given time to reveal herself on her own terms. Time spent in the woods is never wasted. Or as a friend of mine beautifully put it: 'This is not the holiday world of time taken out of real life – this *is* real life. This is a real place and we so need to be in real places.'

Life in modern Britain is often lived at breakneck speed with little regard for where we've come from or where we're going. We are surrounded by so many fleeting distractions, and often placed under such disproportionate pressures to deliver, that it's sometimes easy to forget that the world existed for a very long time before we arrived and will (we can hope) continue to do so long after we're gone. We're simply passing through. And pausing in the moment, sitting still in nature – even when it might seem that not much is happening – can create a deep sense of connection that helps us remember all of this.

By 11 a.m. the first people are percolating down into the woods from the car parks. A young roe buck comes thudding down through the trees behind me to leap over the ditch 30 feet away. He doesn't see me in his haste to escape a group of mountain bikers skidding down the ridge above. The woods now echo to the squeal of bike brakes and the barking of dogs. I'm often amazed by how much human disturbance wildlife can tolerate and foxes are nothing if not adaptable. They're pretty resilient as long as their boundaries aren't crossed and, being so pragmatic, they seem to have us humans well dialled. I hope that these particular young foxes are becoming accustomed to hearing people pass by and will be content to lie low and watch from cover – but it's not really people that are the problem today.

The beginning of the end – in terms of today's filming at least – is a chocolate lab bounding through the middle of the den to the accompaniment of insistent yet consistently ignored shouts and blasts on a distant whistle. I resist the temptation to scare the life out of it as it barrels past me with tongue lolling. Aside from the initial impact, the whole area is now saturated with its scent and that of the last human to touch it. The chances of seeing any foxes today are dwindling rapidly. I feel myself getting angry, but then remember a time many years ago when I was confronted by a forest keeper who threatened to shoot my own dog if I didn't put it back on a lead. I was fourteen at the time and recall the alarm I'd felt when a tall shadow silently detached itself from the trunk of a tree and stepped into the open, rifle in hand. Keepers have obviously worked on their PR since then, but right now he quite honestly has my full sympathy.

The background noise of people passing through the woods is now constant, but I can tell from the direction and flow that most are keeping to the main track. I'm beginning to believe that the cubs may yet still appear when another dog comes bouncing through. This time it's a terrier that knows exactly what to do when faced with a foxhole. It stops to cock its head at an entrance and is about to enter when distant shouts call it away.

The same church-like acoustics that elegantly lift and carry birdsong through the wood now amplify the screams of two young children having a synchronised meltdown. I listen to the mounting desperation in their father's voice as he tries to placate them. I empathise – any parent has been there – but at this point I write off the rest of the day. I stay hidden in my ditch but resign myself to the inevitable and take solace from the tiny birds that are once again visiting the brook to bathe. I watch a blue tit really go for it, fully submerged, its tiny wings whirring hard as they propel the wet bird up to its usual branch to preen.

The low evening sun illuminates the canopy above and the stained-glass glow of leaves melts in reflections on the brook's surface. I've not seen hide nor hair of a fox all day long. The last straw is a couple of would-be festivalgoers pushing their bikes through the deep leaf litter on top of the den. One wears a bikini, the other shorts. They're covered in glitter and both have stereos banging out techno strapped to their panniers. It wouldn't be so bad if they were playing the same track. I pack up and crawl away. No wild fox could tolerate this, and I fully expect the vixen to move her cubs to a quieter site under the cover of darkness tonight.

*

Today has been a bit of a shock to the system. Maybe I've grown too used to having the place to myself. A lot has changed during this past week. The ambiguity over how far people can travel from home is causing confusion and it's obvious that most of those passing through today were from outside the forest. Not that this is bad. Quite the opposite in fact, since the best chance places like the New Forest have of remaining protected is if they are cared for by as many different people from as many different backgrounds as possible. People can care only if they become invested and this can happen only if they are able to develop personal relationships with places through open access. So, it's not a question of whether visitors are local or not (I'm not any more); more a question of how we behave. Loud music and unruly dogs are one thing, but I am growing increasingly nervous about what the rest of the summer might bring.

Sunday 24 May

I'm sitting outside the front of our house watching the sunrise. Light spills over into the valley and shadows pull back beneath the hills. The village is sleeping. Beyond our hedge the road is silent but above me the sky is alive with sound as martins and swallows swoop, mingle and chatter.

Steam rises from my morning tea as I watch El Pijeoto eat his breakfast. Every clang of the feeder is met with muffled clucks of indignation from inside the coop so I let the chickens out. The pigeon fixes me with a beady eye, but doesn't move until chased off by the two hangry hens.

Returning to my bench, I pick up the book serving as a coaster. It's a copy of Eric Ashby's *My Life with Foxes*. It must be more than fifteen years since I last opened it, but Ashby pioneered British wildlife filming, so I figure it's a good place to search for a bit of vulpine-related wisdom. And I'm in need of some. I've been plagued by doubt and uncertainty since my session at the den on Thursday. There's no way the adults could have missed the scent of Labrador when they returned to check on the cubs. Even I'd been able to smell the damn thing as it blundered past, so the ground must have been drenched. But would this have been enough to make the vixen abandon the site, even if this seemed the obvious outcome to me?

It doesn't take long to find what I'm looking for and Eric's words seem uncannily relevant: 'Perhaps most frustratingly of all, long waits at fox earths where there was no action at all for days on end would mean that I was plagued by fears that the vixen had moved her cubs elsewhere – with no way of finding out without disturbing the den, I abandoned many earths only to find out later that they had been occupied all along.'

Eric knew more about foxes than anyone else I know. So to read that he had experienced similar situations makes me think I might have been too quick to accept defeat. I ping Matt a message suggesting he gets back there to watch the den, just in case I've underestimated the vixen's resilience.

Eric didn't just love foxes, he *lived* them. What's more, he was based in the New Forest. In fact, not far from the earth I'm filming stands a bench dedicated to his memory. His film *The Unknown Forest*, broadcast on BBC1 in January 1961 when he was already forty-three years old, immediately established Eric as a household name. In the words of nature writer Richard Mabey, that film 'permanently changed the standards for homegrown wildlife documentaries', to such an extent that 'anyone brought up on modern natural-history films may find it hard to appreciate the impact made by this intimate look into the lives of our native animals'.

Eric went on to make several other films in partnership with BBC Bristol, including the first nature documentary filmed in colour, and his refusal to film tame or captive animals helped establish an ethical code that most of us still strive to adhere to today. The conservationist Sir Peter Scott called him 'The Silent Watcher' and when I look at old photographs taken of Eric at work, it's obvious what a pertinent and fitting epithet this was. What he managed to capture on his antiquated, bulky and noisy 16 mm film camera beggars belief. And the fact that he did it all without the aid of modern light-sensitive film stock or optics, or even resorting to wearing camouflaged clothing, speaks volumes. His knowledge of the animal was so deep, his fieldcraft so good and his imagery so intimate that it's as if he simply stepped through into the animal's world where he was welcomed and accepted as one of their own. No, they don't make them like Eric any more. I wish I'd had the opportunity to learn from him in the field, first-hand.

Eric was in his seventies when my aunt Ishbel took me to meet him at Badger Cottage. He and his wife Eileen ran a fox sanctuary in their back garden. Back then, fox-hunting was still very much part of the forest's identity and there was no shortage of orphaned cubs or injured adults requiring safe haven. I couldn't have been more than fourteen but can vividly remember being invited into one of the enclosures for my first close-up experience of a real live fox. Handing me a chocolate button, Eric told me to hold it in my lips and crouch down. No sooner had I done so than a young vixen bounded over towards me. Before I knew it, she'd leapt onto my back, licked my ear, and snaffled the treat. I still remember the sight of her wrinkled nose and needle-sharp white teeth as she delicately plucked the morsel from my mouth.

Eric then spent an hour giving Ishbel and me a guided tour

of his garden. One of his great film-making triumphs was filming wild badgers underground and by the time we arrived at his shed he had already told us about how it stood on top of an artificial sett that had been built for this very purpose. Wild badgers had moved into this ready-made home and Eric had filmed them through glass panels, gradually increasing the light until these naturally shy animals had grown accustomed to his presence. This kind of filming takes a high degree of skill, a thorough understanding of subject matter and, above all, patience.

Just before we left, Ishbel asked him how relations stood with the local hunts. Back then in the 1980s, three hunts still operated in the forest: Fox, Buck and Hare. A total of 165 meets were held every year. The forest was overhunted and packs of hounds regularly ran riot across the Ashbys' private land. In 1982, a fallow deer was killed in his garden where it had taken refuge, but for Eric the final straw occurred in early 1987 when hounds invaded the sanctity of his artificial badger sett causing the resident family to desert. It was a heavy blow. Not only was Eric in the middle of filming another documentary for the BBC, but the badgers had just started to breed. He was on track to film something truly special, but in the wake of such an intrusion the boars, cubs and heavily pregnant sows all left, never to return.

Wishing to bring an end to such unpleasant disruptions once and for all, Eric applied for an injunction to prevent the hunt trespassing on his land. As Richard Mabey points out, 'it was a brave act made at some personal cost' and unfortunately it attracted the attention of anonymous extremists. Eric received threats. His garage was broken into and a dead fox with its face smashed in was dumped on his front lawn. Despite all this, then in his seventies, Eric persevered in his attempts to film badgers for the BBC. Wild badgers still lived in the natural sett bordering his land, but a year later they too deserted when someone drenched it with chemical animal deterrent.

Soon afterwards, the New Forest Hunt responded to Eric's successful injunction by installing a permanent animal-proof fence along 300 metres of his boundary. This was further extended by an electric fence erected on hunt days. It kept the hounds out but curtailed the natural movement of the many wild animals seeking sanctuary at Badger Cottage. Distastefully, *Hounds* magazine referred to the fence as 'stage one of Stalag Ashby'.

It was Eric's opinion that it was 'the uniformed acceptance of the portrayal of the fox as vermin which allows the barbaric sport of fox-hunting to exist at all'. He hoped that his books and films would help redress the balance by 'portraying the fox as it really is – a rare and beautiful animal no more deserving of cruelty than any other'. A sentiment with which I fully agree.

The hunting of wild animals with dogs was finally banned in 2004, a year after Eric's death at the age of eighty-five. I wish he'd lived to see it.

Tuesday 26 May

I'm in my ditch – waiting. When Matt came here on Friday evening, in the hopes that I might be wrong about the foxes deserting, he'd seen the cubs out playing in their usual spot. He'd also seen the vixen scent-marking. They were all still here, having simply taken the bikes, banging techno and dog visits in their stride. Eating humble pie, I realised I'd allowed my own irritation to cloud my judgement; blaming others for something that hadn't even happened. I am now doubly irritated with myself for not getting back here with my camera sooner.

The midges are murderous today. Clouds of them float in the still air, stinging like finely ground pepper. My head net and gloves keep most of them at bay but inevitably some of them find a way down my neck and up my sleeves. They're so tiny that the merest breeze would keep them away, but this would also carry my scent with it, so on balance I'm grateful for the still conditions. The bracken is a lot taller now, having gone through a growth spurt over the past week and its fronds stack up in dense layers of green through the telephoto lens. The only way I'd know whether the cubs were out and about would be the occasional tremble of a frond, so I make the decision to risk snapping some of the stems in an attempt to open the scene up a little. I daren't leave my scent on the actual den but can certainly make a good dent in the plants closest to me in the foreground. So, creeping forward on my belly through the leaf litter, I spend twenty minutes silently bending and flattening as many stems as I can. Slipping back down into my ditch, I lie low and wait for everything to settle before checking my handiwork through the camera. There's now an open corridor between me and the fox earth, but only time will tell whether the cubs will notice the change or even care.

The day grows hot and the hours ooze forward like the gloop of muddy water and leaves below me. Around midday I'm roused from a reverie by the panicked struggles of a newly fledged coal tit chick caught in my camouflage netting – a short-tailed, ragged-down puff of feathers that manages to get so entangled it takes me a full minute to unpick it. I open my cupped hands and it immediately whirrs and bounces away through the leaf litter.

The cubs finally appear at 5 p.m., but there are only three. Broken Nose and one other are missing. I manage to snatch a few shots, but for the most part they lie low, as if recovering from some sort of ordeal. An incessant cacophony of alarm calls echoes through the woods at dusk. Like a broken dawn chorus, the metallic strikes of blackbirds, robins and wrens join the harsh rattles of jays and mistle thrushes. A discordant wall of sound creeps slowly towards me through the trees then carries on past and I'm left wondering what on earth it was all about. Only a few predators could elicit a response like that and I'm not sure a fox would qualify.

Wednesday 27 May

I'm back with the goshawks and by my reckoning the oldest chick must be about twelve days old now. In the same order as they were laid, goshawk eggs hatch sequentially over the course of a few days. It can take up to six days for them all to hatch and this has obvious implications for the rate of growth within any given brood.

When food is in short supply, any smaller chick unable to compete or defend itself is generally the first to go. In a mixed-sex brood of goshawks, this is usually the smaller male, unless he also happens to be the oldest and therefore been given a head start. There will come a point when he'll nevertheless need to exercise extreme caution in the company of his larger, more domineering sisters. In a power play fit for a medieval royal court, he could well end up on the menu himself.

Siblicide, infanticide (by parents) and even cannibalism (by both parents and chicks) is a well-documented fact of life in goshawk nests. If food is scarce when the chicks are still very young, the smallest may be outcompeted and starve to death via a process of attrition known as 'runting'. If the chicks are subjected to severe food stresses when a little older, the smallest may be killed deliberately by its siblings or even by a parent. Male chicks have been found dead below nests with head injuries from talons and beaks – killed to enable meagre rations to go further. Occasionally, such a victim of 'cainism' may even be fed to its killer by a returning parent. Waste not, want not, I guess, but there's a disturbing slant to such behaviour that can feel unsettling to the human mind – especially at a time when potential food shortages or disruption to supplies have so recently featured in the media ticker tape of the pandemic.

In fact male goshawks seem to live their lives in trepidation of their female counterparts from day one, even into adulthood. I can't say I blame them, having seen the way the females come in to land with talons outstretched, whether the father of her brood is beneath her or not. Again, a fairly universal raptor scenario. I'm reminded of a tree known as 'the shed' growing on the cliffs next to a famous peregrine site in the Wye valley. 'It's where the male goes to hide when his mate's on the nest,' a friend told me. Maybe this tiercel has a 'shed' nearby too.

By 8 a.m. I'm standing directly below my own treetop shed. There's no way I would have risked such a brazen daylight climb when the goshawks were still sitting on eggs, but the chicks have hatched and are now old enough to tolerate being uncovered for a while, so the penalty for flushing a bird isn't quite so high. This 'late' start feels almost decadent, but very welcome after yesterday's long stint with the foxes. From down here the nest looks deserted, but sunrise was three hours ago, so the chicks have probably been fed by now and should be quietly digesting their breakfast. I can also see that more twigs have been added since I was last here. The main landing platform had been looking flat and trampled but now seems fluffed up in a freshly woven springboard of small branches.

Halfway up my ropes, I'm caught out by the returning female, who notices me at the last moment and veers away from the nest. I see her silhouette against the blue sky above as she wheels around the conifer spires, making a strange buzzard-like mewing that I haven't heard before. She doesn't seem aggressive or particularly stressed, but it's a strange sound. I reach the hide, she swoops down through the canopy and lands on the nest, taking no notice as I hoist up the camera.

It's breezy today. The wood is full of air and updraughts that fill my hide, making the sides billow out. I sway gently as sunlight and shadow flicker on the canvas like flame. The female hops up onto a branch directly above the nest and, finally, I get my first proper look at the chicks. There are three. The day is hot despite the breeze and they lie exposed, panting and squinting on the newly raised edge of the nest. Their down is fluffy white, and they struggle to support the weight of their tiny round heads which loll and wobble whenever they stretch out their pencil-thin necks. As with the fox cubs, and the rescued chicks before them, their eyes are slate grey. Likewise, their eyes will gradually change to yellow as they grow. One of my favourite things about goshawks is that their eyes then continue to darken as time goes on. The older the bird, the deeper its eye colour. Which means that our tiercel with the ruby-red eyes is probably eight or nine years old. A venerable age for a wild bird.

This fire in a goshawk's eye is one of their most defining physical traits. It's a feature they share with many other true hawks or Accipiters, such as sparrowhawks, but strangely enough, not with falcons. Theirs tend to be a lot darker. A peregrine, kestrel or hobby's eyes appear almost black from a distance and perhaps this has affected man's relationship with these two very different families. Looking into the eyes of a peregrine can be a calming experience; looking into those of a goshawk can be alarming. Apart from the fact that goshawks can readily dilate and constrict each pupil independently of the other, the contrast of bright yellow iris against its black pin-pricked pupil creates what we might deem a slightly unhinged, maniacal impression. Over time, this has combined with their inherently highly strung nature to feed their psychotic reputation – a reputation that isn't without its own fascination, or even allure, I might add.

The dark eyes of falcons, on the other hand, seem to mimic the open, fully dilated pupils synonymous with human attraction. Maybe we feel a closer affinity with falcons as a result. Maybe on a primal level they appear friendlier, whereas the stark eyes of a goshawk seem to lay their flinty-edged intentions bare. It's a theory. On the other hand, such speculation might be nothing more than the crackpot result of too many hours spent sitting up a tree in solitary confinement.

It's now midday and the chicks are clearly uncomfortable in the heat, dust and flies. One of them has slid back down into the fir-lined cup, but the other two remain in the open where they fidget and jostle. Mum stands sentinel above, stock-still, as only a goshawk can. Her face is split between light and dark, giving her a Janus-like appearance. I see the sun reflected in the black pupil of her left eye and her white cheek feathers kick back highlights that my camera struggles to capture. The tree heaves slowly in the soft cazalty breeze, but she remains perfectly still, staring blankly. Nothing betrays her thoughts or intention.

Goshawks seem to retreat into themselves at times like this and I wonder where they go to in their heads. They can spend hours in these trances, with just the occasional wipe of a nictitating membrane across their eyes to show they haven't been turned to stone. Then suddenly, as if roused from hypnosis, they snap out of it and they're back in the present. Such fascinating creatures.

I'm pulled out of my own daydreams by the hollow shouts of an angry woman echoing through the woods. At first, I think she's below my hide shouting up, but then realise it's coming from where the cars are parked. Worried for Matt, I send him a text. The shouting continues unabated and although I can't make out specific words, there's an indignant, rising tone that makes me want to head down to check Matt's OK. Just as I'm about to risk breaking cover, my phone pings: 'Angry local. Horse rider wanting to know what the "bloody hell" we're doing here in her woods.' Her woods. I guess I'm not the only one becoming anxious in the face of an influx of people.

Thursday 28 May

The three remaining cubs are above ground by 10 a.m. Their tiny bodies lie perfectly camouflaged in the leaf litter, soaking up the morning's warmth with only the flick of an ear to draw the eye. Their juvenile guard hairs stand erect above their woolly underfur, 'like wild oat in a field of wheat' as scientist Gwynn Lloyd so eloquently put it. This gives them the vaguely fuzzy outline that helps them blend in with their soft natural surroundings.

Despite the spring sunshine there's a slightly subdued feel to the scene, and I can't help wondering where the other two cubs are. Vixens sometimes discard weaker cubs by abandoning them to die on the edge of home territory. Although Broken Nose was certainly the runt of the litter, he was nevertheless the liveliest and most active. In any case, this wouldn't account for the disappearance of the other cub, who was to all appearances perfectly healthy and of good size. Trying to second-guess such mysteries can be a futile exercise. There are of course multitudes of reasons why young cubs might disappear. Aside from straightforward accidents or predation, their mother might have simply moved them to a different earth nearby, where they continue to thrive out of sight, but I've been in a similar situation before and, in that case, it didn't have a happy ending.

Ten years ago, I worked on a friend's film for the BBC about the Lost Gardens of Heligan down in Cornwall. All the usual characters were there: badgers, barn owls, roe deer, and one of our main characters was a vixen raising her cubs in the grounds of a manor house. I spent many happy evenings filming them at play on the manicured rhododendron-fringed lawn. Once again, the smaller of the two was the more playful. An endearing little male who took great delight in ambushing his older brother and rolling down the grassy banks with him. Then one evening there was only one cub with the vixen. I was sleeping in an old building elsewhere on the estate and on opening the door the following morning I was greeted by the pungent odour of fox musk and the unexpected flurry of a herring gull taking flight a few feet away. It had been feeding on the small corpse of the younger fox cub. The air was acrid and heavy, and the cub's front leg and shoulder were missing. A long tendril of sinew had been tugged out of the hole by the seagull, but the bird certainly hadn't killed it. The only conclusion I could come to was that it was either run down and killed by a dog or had been deliberately shot and dumped there. The fact that the mutilated body had an entire leg missing and had ended up right outside my front door immediately played on my mind. I mused on this mysterious, tragic turn of events for weeks afterwards before finally deciding it was best to let it go.

The inescapable fact is that mortality rates in wild foxes can be very high. It's one of the reasons vixens give birth to so many cubs. The odds are stacked against the tiny scamps from the very moment they set foot above ground. It's exactly the same for songbirds, mice, rabbits and all the other creatures. We tend to celebrate springtime as a joyous period of awakening, fecundity and new beginnings: the season of life. And so it is. But it's easy to forget that springtime is defined by death just as much. The pressure placed on parents to bring back a never-ending supply of food results in nothing short of a seasonal killing spree. We just don't tend to see it, since it usually happens out of sight in hedgerows, bushes or simply in the very early morning or late at night when we're not around. For every wood pigeon brazenly killed by a sparrowhawk in the garden, a thousand more die in secret. Life and death are partners in an eternal dance that ebbs and flows. The balance must tip in life's favour for a species to survive, but the sheer level of attrition among the newly fledged and weaned can be shocking, especially when scrutinised through a 1,000 mm lens. It can be heart-rending to film sometimes, but one animal's sacrifice often represents survival for several others.

Just before I left Kenya in March, one of the cheetah cubs I was filming was killed by a lion. Like the goshawk that predates the nest of a fellow raptor, it somehow seems wrong for one big cat to kill another, but this is just the way it is and life sets out to win at all costs.

Watching the remaining three cubs dozing, I accept that I'll never know what's happened to Broken Nose and his sibling. In a flash of irony, I recall reading about one unfortunate cub that turned up on a goshawk nest. A large female gos is certainly capable of this, although I don't think for a second that this was the fate of Broken Nose. Not when there is so much other prey around.

As the afternoon goes on and the sundial shadows of bracken creep across the sleeping cubs, I am on the edge of drifting off myself when I'm startled awake by a tiny avalanche of dry soil. A wood mouse scurries along the bank before disappearing down a hole. A few seconds later it emerges on a patch of grey-green bryophytes down at the water. Springing over the ditch, it scampers up the opposite bank, sits on a moss-covered log and turns to regard me with unfeasibly large black eyes. A moment later it's gone, its yellow fur melting into the leaf litter. Two minutes later its grey arboreal cousin, huge by comparison, hops through the leaves to stare at me, flicking and whirling its bushy tail like a maniac. It clearly doesn't think I should be there and huffs angrily to make a point. I keep still and it comes close enough for me to see how muscly its shoulders are. A heroic bound and it's 10 feet up a nearby oak, staring at me with barely concealed indignation.

By 6 p.m. the stems of trees are burnished bronze by the sun, but the cubs still haven't moved. Only the occasional rabbit-dream twitch of a foot shows they're even still breathing. A slow day. Colour seeps out of the scene and the woodland takes on a bottle-green hue as dusk gently settles.

And then, just before light is lost altogether, there's a twitch of bracken on the ridge and I press record on the camera. Barely a second later a flash of white announces the arrival of the dog fox. He is trotting forward, weaving his way through the bracken stems down towards the cubs with his tail swinging behind as if brushing his tracks. He carries food in his mouth, a bird of some kind, and his pricked ears and loping gait speak of confidence. The vixen emerges from behind an oak tree to greet him. Only now do I realise she's been there all day, lying up close to her offspring. The dog becomes rueful and nuzzles her with flattened ears and sloping shoulders. There's even a wag of the tail as he continues past in search of the cubs. His posture is now low and non-threatening as he approaches the den entrance and, as he turns side on, I get a proper look at his tail. It's fabulously bushy, flecked with grey and tipped with black. Combined with his black stockings and starched-white breast, he presents the picture-book ideal of a rural fox in rudest health. A second later, all dignity is lost as he's bundled to the ground by three pups eager for food and attention. They are beside themselves with excitement, running through the bracken, tripping up their dad and generally not knowing what to do with themselves. The vixen is content to stand by and watch as the food is dropped in front of the only cub now not racing around like a lunatic. It immediately tucks in, adhering to the age-old fox code of first come, first served. The dog fox disappears again in a sway of bracken; the cubs hide, lying low to eat, and the vixen retreats behind her tree. The stage is empty once again. The scene lasted barely two minutes, but every second of it was worth the twelve-hour wait.

I give it another ten minutes, then start quietly packing before it gets fully dark. Then, just as I go to turn off the camera, a herd of twelve fallow ghost in from the shadows to the right. Picking their way slowly through the bracken between the oaks, they are less than 30 feet away but show no sign of having scented or seen me. I study them through the lens. It's a group of does led by an older female, the matriarch of what is probably an extended family of mothers, sisters and daughters. They walk slowly in single file, white tails flicking in the twilight. One by one they follow the curve of a trail hidden beneath the bracken. The whole forest is criss-crossed by narrow deer paths that have been used since time immemorial. Matriarchs use them to lead others to food, shelter and traditional breeding sites. These animals are walking in the footsteps of a long line of ancestors stretching back centuries, following song-lines woven through a landscape layered with history and meaning.

As they turn towards me, I realise that several are heavily pregnant. Their swollen flanks roll gently from side to side with each delicate step. Within the next few weeks, each expectant mother will go off to give birth on her own in some secret place. She will keep a low profile and suckle her tiny speckled fawn until it is old enough to run with the herd later in summer.

The last doe to pass seems to sense something strange. She can't pinpoint me and stares off to the right, but I see her lick her nose and crane her neck up to catch any passing scent. Her tail isn't raised, so she isn't alarmed, just suspicious. Lowering her head, she pretends to graze before lifting it back up with a jerk a second later. She's trying to catch me out. Pretending to be feeding in the hope that any would-be predator will be duped into breaking cover. It's an age-old game played by deer all over the world. Her huge dark eyes glisten as her lower jaw chews the cud thoughtfully. She breaks into a half-hearted trot to catch up with the herd and within minutes the tall oaks are engulfed by shadows. It's time for me to leave.

Friday 29 May

Up at 3 a.m. On the road by 3.45, a van overtaking mine so quickly that my own wobbles in its wake. Coming into Lyndhurst, I'm sad but not surprised to see a freshly killed fox on the side of the road. First casualty of the day.

By 4.55 – sunrise – I'm quietly ensconced in my hide where it's always twilight. It's a cold start but I don't mind the boreal chill in my bones. It seems somehow fitting. The chicks are lying low in the nest, their mother standing off to the side in a trance once more.

Suddenly she flicks her head to the left, her eyes fix on something unseen and her shoulders hunch forward into a predatory stance. She opens her beak as if she's going to call and a split second later the tiercel appears on the edge of the nest, spilling the air from his wings with an almost delicate flutter. He carries the body of a songbird in his right foot. Folding his wings, he stands to face the female, who shuffles forward until they are almost beak to beak like sculpted bookends. Once again, I'm startled by how fey he is. While her powerful form suggests strength and endurance, his lighter profile speaks of speed and stealth.

He takes a step forward to stare at the chicks before plucking at the small carcass beneath his feet. The female stays rooted, but opens her beak to emit a soft keening sound as if begging. I had expected the tiercel to have left by now, but to my surprise he starts tearing up the food himself. All three chicks immediately lurch towards him, their tiny crooked wings flapping up and down to keep balance.

This is the first time I've seen the male goshawk feed his young and the first decent view I've had of all three chicks moving around. I let the camera roll as I follow the behaviour through the lens. Holding a small strip of flesh in his hooked beak, he leans down, tilting his head to one side. The nearest chick tracks every movement with huge grey eyes until it can't stand the anticipation any longer and makes a lunge, but the male fumbles and drops the meat before it can be taken. The chick looks confused and the male tries again, this time offering a wisp of silky feather that is eagerly snatched. The female doesn't seem to like what's happening and edges towards him, mewing softly. This time he gets it right and manages to place a large strip of flesh directly into the chick's waiting beak. Nevertheless, the female has run out of patience and stepping over the chicks she gently but firmly displaces the male, taking possession of the food. He just has time to pluck a morsel for himself before he's nudged off the nest and the female gets down to business, tearing at the body with decisive tugs. Her feather-plated back is towards me and I can see the tips of her new primary feathers crossed over the base of her tail like scissor blades. The old primaries were the first feathers to be dropped as part of her annual moult and these fresh replacements aren't yet fully grown. They are darker than the older weather-worn feathers on her back and are ridged, giving their silver edges a serrated appearance. Her coverts flash as she leans forward. Bright white and wispy, they curl around the base of her stiff tail like vortices in cold air.

The departed male calls in the distance and the female looks up for a moment, revealing two white spots on the back of her head. All adult goshawks have these, although I'm not sure what they are for. They remind me of the white dabs on the back of tiger ears, a mark for cubs to follow while being taught how to hunt in the long grass. Maybe they have a similar function in goshawks or simply form part of the elaborate courtship displays in spring. They may even be a form of mimicry, like the distracting eyespots known as ocelli seen on the wings of butterflies. Fake eyes can be useful in warding off attacks from predators duped into believing they've been spotted. Goshawks are ambush predators, spending long periods of time perched in one place. Having eyes in the back of your head could be useful in helping to prevent the ambusher from becoming the ambushed – a little like the human masks that some woodcutters wear on the back of their heads in tiger country. It also occurs to me that raptors are at their most vulnerable while feeding on the ground, so perhaps the eyespots help deter opportunists dropping down from above. The sudden flash of these marks as the feeding goshawk jerks its head up from a meal might be enough to confuse or deflect.

There's no doubt that goshawks are top avian predators throughout their range, but whether they occupy the rarefied niche of *apex* predator in any given ecosystem depends very much on which other raptors they share their territory with. There are two dozen or so species of true goshawk in the world, many of which live in the tropics, but the only species to be found in Britain is *Accipiter gentilis*, the northern goshawk. In fact, this is the only species of goshawk to be found resident throughout the upper northern hemisphere, where it – and its various subspecies – occupies a circumpolar distribution from Kamchatka in the east, right across northern Eurasia and Canada, to Alaska in the west.

I'm sure that many species of eagle would take great delight in killing a goshawk, but for the most part boreal eagles don't tend to hunt in the same densely forested habitats. Unfortunately for goshawks, eagles aren't the only would-be predators out there in the trees. Some of the larger species of owl are perfectly capable of taking a goshawk, while it's asleep or otherwise distracted. The apex avian predator in many Eurasian forests is the European eagle owl, the largest species of owl on the planet. These supreme predators can weigh well over 4 kilograms – twice as much as a large female gos – and have a wingspan approaching two metres. In North America the equivalent niche is occupied by the great horned owl. Both species are goshawk killers and both favour hunting in the crepuscular hours around dawn and sunset, a time when colour drains from the landscape to be replaced by greyscale. At times like these, it is the contrast between shades that counts, and white eyespots on a dark-grey background could prove startlingly effective.

While there aren't any eagle owls in the New Forest, goshawk DNA doesn't know this, so unless the eyespots are actively selected against over time, there's a good chance they'll simply remain as a neutral morphological feature. A memory of ancient evolutionary pressures. On the other hand, there might well come a time when such features become useful once again.

Whether we like it or not, eagle owls are currently making a concerted effort to recolonise Britain. I use the term 'recolonise' deliberately, for there seems little doubt in my mind that they were formerly part of our island's ecosystem. I am also acutely aware that the issue of whether *Bubo bubo* should be considered native or not is a contentious issue for many, including the RSPB. The debate also revolves around whether the birds currently living in the wild are escapees or natural immigrants from mainland Europe. The North Sea is a hop, skip and a jump for a bird with a six-foot wingspan, after all. Personally, I suspect the truth is neither one nor the other, more a mixture of both. Either way, in these modern days of social media, there is no refuting the established presence of these birds in the British landscape. The internet is awash with images of them, mostly from the north of England near to where the first-known pair of recent times began breeding in the late 1990s. Current population estimates vary wildly, but there seem to be around two dozen pairs nesting out there. Eagle owls are nevertheless currently classed as 'non-native' by both the British Ornithologists' Union and RSPB.

All of this seems somehow reminiscent of the goshawk's own history. Until very recently, goshawks were considered officially extinct in Britain, our native population having been decimated by persecution in the 1800s. Rumours of feral birds breeding in the wild began to grow in the mid-twentieth century and it was tacitly acknowledged that these had descended from captive birds that had either escaped or been deliberately set loose.

Given the innately nervous and fiery disposition of the goshawk, it was perhaps inevitable that the occasional falconry bird would be lost while out hunting. Indeed, Lascelles even admits to losing his beloved hawk Shelagh 'owing to a clumsy blunder with rotten tackle' way back in the late 1800s, more than a century before goshawks began to recolonise the New Forest in earnest.

There are also tales of twentieth-century austringers (falconers who fly hawks) purposefully importing two birds from the Continent with the express intention of keeping one while hacking the other back to the wild. To 'hack' is an old Elizabethan falconry term describing the process of preparing captive chicks for a life on their own in a suitable habitat, a form of rewilding in both senses of the word. Goshawks are extremely independent birds that generally respond well to being hacked and the wild British population seems to have grown quickly as a result. It nevertheless took the establishment a while to acknowledge and accept that the goshawk was back to stay. As Leslie Brown put it in his 1976 edition of *British Birds of Prey*: 'There seems to be little point in writing at length about the goshawk in Britain since, although there have been persistent rumours of increased if sporadic breeding in this country for about the last three decades, factual evidence for this in the published literature is negligible.'

But what's even more interesting to me is what he then goes on to say: 'whether any or all of these are genuine wild birds that have re-established themselves here from outside, or whether they are all falconers' escapes, or the offspring of falconer's escapes is unknown. *However, I personally can see no reason why the first-generation offspring of falconers' escapes should not be regarded as wild British goshawks from now on, if in fact they exist* [my italics].'

It's a sentiment I tend to agree with, but a contentious one nonetheless. Brown was a world authority on raptors and clearly very passionate about their place in our countryside, but even he seems to have been driven to distraction by the apparent ambivalence (if not downright apathy) displayed towards any attempt to secure official recognition and protection for the returning goshawk: 'It seems to me tragic, and symptomatic of the utterly unrealistic attitude of the

British towards their birds of prey that such a situation can exist at all.'

Things have obviously changed a great deal for goshawks over the past fifty years, and I like to think that Brown would heartily approve of the high level of protection now afforded to this 'magnificent raptor'. This isn't to say that persecution doesn't still occur, however; rather it means that those caught doing it face potential consequences. Once again, I choose my words carefully here since such legal action brought against those charged with killing protected British raptors has proved to be a rather hit-or-miss affair over recent years. Also, I suspect that for every high-profile golden eagle or hen harrier poisoned or shot on an exposed mountainside or grouse moor, many more goshawks meet their end without anyone ever knowing. The snap of a trap or thump of a .410 are easily missed in the woods.

Goshawks returned to the New Forest in 2000, with the first official nest found by Andy in 2001. They were the first pair to breed here for well over a century. Andy suspects they were descended from three or four birds released by falconers over the border in south Wiltshire a few years previous to that. In the space of nineteen short years, goshawks have effectively repopulated the entire area and the forest is now at saturation, with forty-five active breeding territories. It's as if they've sat down with a campaign map and systematically carved up the landscape between them. This has been confirmed by the sharp rise in goshawks dispersing beyond the forest borders. Breeding birds have now been found as far away as west Dorset and north Hampshire. The forest is a stronghold and its population represents a significant proportion of the 400 or so pairs now thought to be breeding in the wild in the UK. These are still relatively low densities overall, but it's clear that the goshawk is here to stay.

I see the undeniable return of the goshawk and the tentative return of the eagle owl as stamps of approval. Animals don't linger in habitats unable to sustain them and although eagle owls naturally disappeared from Britain several millennia before the persecuted goshawk, the fact that both birds seem to be successfully reoccupying extant niches should perhaps encourage us to welcome back other native species if and when they similarly reappear.

Nature doesn't always need grand gestures of political support via wide-ranging environmental reforms that spend years navigating the corridors of power before they are implemented, if at all. I believe that a little space goes a long way and sometimes all we really need to do is take a step back to let nature do its thing. A helping hand is sometimes welcome, but to think that nature needs constant micromanaging smacks of hubris and to my mind simply reflects our generally elevated sense of self-importance.

Which brings me around to my initial hopes for lockdown. At its simplest, nature responds well to a little less pressure from us at a local level. In fact, I'd even go as far as to say that this is all some species require. It is nature's ability to help itself, to survive *in spite* of us in fact, that gives me tentative hope, not only for the future of our own nation's wildlife, but also for that of the planet as a whole. As long as we can stave off wholesale habitat destruction – and I realise this is where large political gestures often *have* to come in – local wildlife can become surprisingly tolerant of its human neighbours.

I once filmed chimpanzees at a place called Bossou. Sitting in the shadow of the Nimba mountains on the border of Guinea and Liberia, West Africa, it is home to a wild population of apes that have learnt to use seven different natural tools when foraging for food. Expecting a remote tract of rainforest, I was amazed to find them free-ranging through a loose conglomeration of villages,

140

farms and fragmented jungle. I'm not saying that this was ideal, just that it amazed me to see how resilient and adaptable the chimps were and how tolerant the local people were of them. We often assume that there has to be an all-or-nothing approach to wildlife conservation and that landscapes have to be protected in pristine condition for them to remain of any value. I would argue, however, that this perspective sometimes proves counterproductive since the challenge of keeping things in a pristine state might prove so intimidating, and ultimately so unachievable, that a sense of futility creeps in and all hope is lost – when it need not be. I'm not saying that all species can cope with heavy environmental impacts, or that we shouldn't strive to hold on to as much wilderness as possible, just that if and when the obstacles to complete preservation prove insurmountable, other options should remain on the table. We should take heart from the fact that nature is often capable of meeting us halfway and use this as motivation for doing our own bit, however modest this may be.

I'd love to see eagle owls back down in the forest one day, but for now this pair of New Forest goshawks are at the top of the avian food chain and, back on the nest, the female demonstrates this in visceral fashion by continuing to tear up the carcass pinned beneath her talons.

Unlike the male, the female knows exactly what she's doing and at last I have a great view of all three chicks jostling and craning up their skinny necks to receive the food. They seem to be taking turns in a surprisingly polite fashion and the little round heads pop up one after the other like periscopes. All are covered in a tightly woven suit of velvety down, almost like a fleece. Halos of longer wispy down float above their lollipop heads like punk hairdos and their brow ridges are well developed, as are the yellow ceres at the base of their beaks.

Their beaks are tiny and needle-like and can already grip and tear, although most of the food goes down whole in one gulp. A few feathers are no bad thing, but for now the female takes great care to feed them only the choicest morsels, lest they choke. It takes time, but when she's finally finished, the chicks each have bulging crops and seem to enter a kind of meat coma, collapsing on top of each other in a spaced-out heap in the bottom of the nest.

The feed should tide them over for the rest of the day, but at this age their metabolism is extremely fast, and their tiny stomachs can hold only so much. This is where their crops come in. Not all birds have them, but all raptors do, and it is a neat solution to the problem of having to eat as much as possible in one sitting before the food source is either taken away or has to be abandoned. 'One thing drives out another', as the saying goes, and the crop's contents will gradually move down into the stomach for digestion once the stomach's previous contents have moved along. Even at this tender age, the chicks are meticulously clean in their toilet habits and eject their mutes with enough force to catapult the bacteria-filled mess clear over the side of the nest. With staphylococcus, streptococcus and salmonella all at play in there, it's an unpleasant cocktail of microbes that you wouldn't want lying around to fester.

It's always a little nerve-racking, watching a chick that can't walk properly (let alone fly) shuffle and wobble its way to the edge of a 50-foot precipice before turning around, leaning forward, and firing a stream of white gloop into space. It can also be quite funny, especially when it's done with such force that the recoil knocks them forward onto their face back into the nest, but perhaps that's just the schoolboy in me.

Once the chicks are old enough to feed themselves, their mother will leave them to it with the freshly caught animal and, in direct competition with each other, the focus will be on eating as much as possible, as quickly as possible. As a result, the chicks will also ingest large amounts of extra material of low nutritional value: feathers, fur, bone, teeth and all. Such by-products are not wholly without use, however, and collect in the gizzard where they form pellets that help scour out the digestive tract when later regurgitated. Such pellets are an excellent way to discern what a particular bird has been eating and anything from bird rings to mole teeth show up in goshawk casts.

The chicks are now motionless and their mother has once again retreated to her gargoyle perch above. The hours slide past slowly in the heat and the nest attracts the attention of bluebottles that fly in lazy circles above the chicks.

A few hours later, the male returns with another kill. This time it's a lanky scrap of a thing. All legs, with a pointed bill and white cheeks reminiscent of a young woodpecker. There's barely any meat on it and rather than tearing up the carcass, the female seems content to stand still for the next hour. Eventually she reaches down into the nest to retrieve a scrap of bloodied bone and flies off with it in her beak. A small, round head pops up as she leaves, its white down smeared with scarlet that dulls to brown in the sun. It looks as if the chicks are already starting to feed themselves and that they've spent the past hour tugging at the tiny corpse in the bottom of the nest. Judging by the bloody scraps removed by the female, they didn't leave much, and I wonder whether the tiercel is deliberately targeting small, tender fledglings to help them through this transitional period from being fed to feeding themselves.

It's been an excellent morning and so I decide to spend the afternoon with the foxes. It's only a matter of time until the bracken gets too high to see them or the remaining cubs disappear like their two siblings. I wait until the nest is fully settled, with the chicks asleep and their mother on guard, before radioing Matt. By noon we're driving out of the wood.

It's two weeks since lockdown ended and two weeks since I was last out in the wider forest at this time of day. My first warning that things might have changed here too comes as I get out of my van to unlock the forest gate. Two men on mountain bikes appear from nowhere and swerve between me and the van to cut through the opening. They are gone in a skid of gravel, but my nose is filled with the pungent odour of sweat and deodorant. After seven hours on my own in the trees, I can even smell the shower gel they used that morning – a confused swirl of stale chemicals that almost makes me gag. I'm sure that I smell no better, but after so many hours spent in the wood I've been given an insight into what our species must smell like to a fox or a deer. It's an unwelcome reminder that however much I might try to enter an animal's world, there will always be a gulf between us.

On top of this, I'm angered by the fact that after weeks of wearing gloves to open gates I've just been casually exposed to a slipstream of droplets exhaled by a hyperventilating stranger. It's a thought that brings me up short. Not because of any real fear of infection, more to do with how divisive the pandemic is becoming. We are all in this together, but the knowledge that any person we meet could be carrying a dangerous virus is sobering. It doesn't take long for suspicion and fear to become corrosive, so telling myself not to let paranoia overrule common sense, I get back in the van and drive up the track.

Within five seconds of joining the main road I wish I'd stayed in my tree.

The roads are carnage. I've never seen the forest like it. Torrents of cyclists; cars abandoned on the verges; dogs running everywhere and a wagon train of campervans in the lay-bys. The car park at Janesmoor Pond looks like a festival with everyone parked cheek to jowl. Rows of awnings have been erected, patches of grass claimed and two police officers try to disperse what I can only describe as a baying mob intent on buying ice cream. I watch someone use their bumper to push a wild-eyed pony off the road and on the edge of the tinder-dry heath a disposable barbecue sends its oily smoke up into the clear sky.

It's a wake-up call. It was easy to be lulled into believing that we'd used the lockdown period to reflect and learn from what was going on in the wider world. I'd assumed that once restrictions began lifting, we would act with a little more balance, decorum and humility. Once again I find myself filled with doubt and shame. The forest has always been busy in the sunshine, but what I'm witnessing in this post-lockdown release is an inundation that puts me on edge and makes me nervous for reasons I don't fully understand. I remind myself it's everyone's right to get out now they can, but the old New Forest is under more pressure now than it has ever been during its thousand-year history.

Sandwiched between Bournemouth to the west and Southampton to the east, more than a million people live on its doorstep. London is only an hour and a half away. Way back in 1880, the author of *The New Forest: Its History and Scenery*, John Wise, seems to have been looking through a crystal ball when he wrote, 'The time will someday arrive when, as England becomes more and more overcrowded, – as each heath and common are swallowed up, – the New Forest will be as much a necessity to the country as the parks are now in London.'

Prophetic words, but part of me wants to push back and fight for the forest's right to remain wild at heart. To treat this landscape as a domesticated parkland seems to me a crying shame, as does the notion that this level of disturbance might be another premonition of an everyday future.

The heightened activity I'm seeing in the forest today, however, smacks of something deeper than people simply letting off steam after being cooped up for so long. It's more visceral and unsettling than that, akin to an expression of a country's deep-seated frustration – even anger. There's little joy to be had from competing to buy an ice lolly or sitting in a heated jam waiting for a parking space and it feels frantic, forced and depressingly symptomatic of a nation at the end of its tether, desperate to reclaim control. Ultimately I guess people have nowhere else to go. I've been one of the lucky few these past weeks, but as the realisation dawns that we could now be entering a period when such restrictions on movement and foreign travel become an unfortunate fact of life, I have a feeling that today may represent the Ghost of the Forest Yet to Come.

The narrow hedge-lined roads through Fritham are choked and the main car park there rammed too. Even the gated track into the woods is busy with a steady stream of bikes and rucksacks, so I put my hazards on and crawl along at walking pace while people scowl at me and make a big show of putting their dogs on leads. It seems that we've all had a stressful time getting into the forest today and several older locals glare at me as if to ask, 'Who the hell are you?' Fair enough, I suppose, but my relief at finally reaching my destination is immediate. Crawling beneath the fir trees with my camera, I take a few minutes to get back in sync with the forest before setting up my kit. I never thought that my happy place would be a ditch, but there you are.

By early afternoon, two cubs are out sunning themselves in a small glade amid the bracken. I can hear how busy it is out there on

146

the wooded tracks and trails, so I don't hold out much hope of seeing the adults this evening, but 80 per cent of success is just showing up, and foxes are nothing if not predictably unpredictable.

Unfortunately, though, on this occasion I'm proved correct, but just as I'm packing up for the evening, two ponies saunter down to the brook to drink. Kicking their way through the drifts of leaves, they wake the fox cubs, who sit up for a moment, then collapse once again in a heap. The ponies make the low, soft, contact noises made by all large herd animals. It's marvellous to see how much like wild animals the ponies are when encountered on their own in the woods, far away from picnics and car parks.

I wait until it's fully dark before re-emerging to face the drive home. Things have quietened down and all that remains of the day's chaos is a lot of windswept litter, scorch marks on the grass and empty cans of lager. There's a forlorn though slightly crazed air about the scene and it's hard not to feel ashamed of my own kind.

For me, today's issues transcend the question of how far people have travelled to make it here. I've seen local farmers set fire to tractor tyres in nature reserves and kids from inner cities sit in rapt silence while watching a blue tit, so I don't believe that respect for the environment relies on living close to a national park. In fact, the opposite is very often true since we seem to take the familiar for granted. Nevertheless, if we cannot collectively behave with restraint, continuing to place our own immediate needs above everything else, then there can be only one outcome.

Saturday 30 May

I'm excited. This is my first day filming a curlew nest. Andy left me a message yesterday to say one of those he showed me is still active, although he suspects the eggs may be infertile, a situation he'd been anticipating given that the male's injured leg made it hard for him to mate. Incubation lasts four weeks, so they should have hatched by now. There's no way of telling how long the birds will stay with an infertile clutch, but I guess they'll keep going until the nest is either predated, or it becomes so obvious that they abandon it. It's painful to watch. Curlews have a hard enough time as it is, but to have kept the nest successfully hidden and guarded for the past month only to discover that the eggs will never hatch is a bitter end to their hopes of breeding this year.

The nest is in a wild corner of the forest, a forgotten enclave of ancient landscape that is remote enough not to draw attention to the birds I'm filming. The other curlew site that I visited with Andy was a little easier to reach, with more secluded cover, but it's been predated. Indeed, it wasn't very far from the fox earth I'm filming, an irony that isn't lost on me. Crows will also take eggs, or rather they will break them open in situ to eat the yolk or the developing embryos. The forest also supports around twenty breeding pairs of ravens and these are large enough to drive away adult curlews to get at their chicks.

With yesterday's Hieronymus Bosch vision of picnicking hell still fresh in my mind, I made an early start and by 7 a.m. was settled in position. My filming hide stands on dry, sandy soil with its back to a gorse bush, but the ground in front shelves gently down into one of the forest's valley mires, a wide area of ancient bog that's changed little since it first came into being 11,000 years ago. It's a superb spot with a great view of the transitional zone from heath to wetland. Heather giving way to tussock grass, which in turn gives way to myrtle and bog cotton in the valley bottom.

I've set my camera back from the hide's front opening. The dark interior conceals the lens, allowing me to follow movement without startling the birds. Unlike the goshawks, the curlew I've come to film haven't had time to get used to the hide. Matt and I carried it in this morning in a twenty-minute trek across the heath to the accompaniment of skylarks and stonechats. He stayed with me while I got set up and comfortable inside.

The male bird came wheeling and fluttering back in within minutes of Matt walking away, the still morning air brimming with his plaintive, undulating call. Having landed in the long yellow grass to my right, he stood sentinel for a while before cautiously zigzagging his way back to the nest – careful not to leave a direct trail for unfriendly eyes to follow. Andy was right, this was the same male we saw limping at the edge of the pond when we visited two weeks ago. His injury seemed even more acute in the long grass and he can't move fast, but on the other hand he shouldn't really need to and what elegance he lacks on the ground is more than made up for by his seamless reeling and gliding across the sky. I had a rough idea of where the nest was meant to be, but it was only when he got close and the female stood up to swap over that I could pinpoint the exact spot. She'd been there the whole time, hiding in plain sight. Ruffling feathers as if shrugging off her camouflage, she tiptoed elegantly away before leaning forward and springing into the air in a flurry of powerful wings. I panned the lens back to the nest, but there was no sign of the male and it took ten minutes of pulling focus through every blade of grass to finally relocate him. He's now been on the eggs for the last couple of hours. I presume his mate is still feeding on the edge of the marsh somewhere.

The sound of a distant motorbike going up through the gears makes me realise how peaceful the morning has been so far. For the past couple of hours, I've heard nothing but the breeze whistling through the gorse, a flint-knapping stonechat and another distant curlew. I know that the rest of the forest will be heaving today, but out here between the heather and the sky, all is as it should be and probably always has been. It's still quite early, not quite nine yet, but the air is already beginning to melt and wobble in the sunshine. It promises to be a hot day, but for now the gentle breeze blowing through the front of my hide is refreshing.

As filming hides go, this one is positively salubrious. It's a lot larger than my tree hide, and drier than my foxhole spot, so not only can I stretch out my legs, but I've taken off my boots and don't have to worry about wet feet – or motion sickness for that matter. Noise isn't so much of an issue either. The expansive silence of the heath seems to smother sound. A welcome change from the hushed expectant stillness of the goshawk wood in which every sound promises to startle the gos. Curlew are certainly twitchy birds, but I'm further away than when in the tree platform and there's a lot more going on out on the heath to help dilute and soften the effect of my presence, though the only thing currently moving in my camera's crosshair is the warm, liquid air undulating languidly like oil on water. Settling in for a long wait, I pour myself a cup of tea and enjoy the sensation of the coarse heather on my bare feet.

By lunchtime the air is too full of heat haze for my lens to focus sharply. The light is burnt out and flat and there's no texture in the landscape. Not that the birds have done much to film anyway. I wonder how they don't bake out there, tied to the nest with no cover, but feathers are just as good at keeping birds cool as they are at retaining warmth, so I guess that like me they're simply sitting it out in anticipation of some respite later this afternoon. I notice that the male has got his beak open though and from the rapid movement of his short, spiky tongue, I can see he's panting.

The morning slides slowly on. I eat a can of cold baked beans for lunch and watch a solitary crow lurch through the haze. Another one sits in the branches of a dead pine off to the right. There's always something watching, even when all seems quiet. Corvids spend a lot of time looking for patterns in other birds. Any repetitive routine that might indicate the presence of a nest to rob, and it occurs to me that it's not just the direct threat of disturbance that the curlews are up against. The panicked flight and alarm call of a curlew in distress from human or canine interference rarely go unnoticed and if a crow can use the diversion to sneak in under the radar, then it's all over and the curlew's breeding plans are dashed for another year.

Bikes and dog walkers aside, a curlew's normal defence is to sit as tight as possible and rely on excellent camouflage to keep themselves and their nest concealed. At first glance, the curlew's plumage looks deceptively random and uninspiring, a drab collage of white, beige and brown flecks. Seen in context, it has the magical ability to render the bird invisible. It works well on a winter seashore, but where it really comes into its own is in long, tussocky grass, the no man's land of its preferred breeding habitat between moor and marsh. Exactly what we have here. The delicate shading of its feathers accentuates texture and mimics the vertical shadows and highlights of long grass. What initially looked like a rushed paint job is actually an exquisite cloak. Even the conspicuously long, downward-curved bill seems to mirror the slender, curled-over grass stems surrounding their nests.

Such perfect plumage isn't the product of random chance but has been actively selected for millions of years. Every bird that escapes predation passes its adaptations down to the next generation and, by subtle nudges and tweaks to the base code, the camouflage of plumage, behaviour and movement is honed and perfected. Some creatures such as mountain hares, ptarmigan and Arctic foxes regularly change their hue to bring it in line with seasonal colour, though climate change is beginning to thwart these carefully calibrated adaptations. Others, including chameleons and octopuses, can alter the pigment and even the texture of their skin within seconds. Still, the majority of animals spend their whole lives wearing the same suit, certainly from adulthood onwards. Their designs can be incredibly sophisticated, such as the extreme mimicry displayed by many insects, or deceptively simple, such as the counter-shading seen on predatory fish. For many, once manifest, these designs are habitat-specific and if they are driven out of that habitat, or their habitat is destroyed, then selective pressure will be high, and the race will be on to adapt and conform to their new surroundings. Natural selection is a truly incredible process. How external influences can mould, tame and favour internal mutations is simply mind-boggling, in both complexity and simplicity. So when I look through the camera at the cryptic cloak of a curlew, I see just as much beauty as I see in the dappled coat of a leopard or the burning stripes of a tiger.

Every encounter with this bird is special and I'm sure many people have their own favourite memories. I've watched flocks of them rise into the winter moonlight over the Bristol Channel, seen them probe the sand next to seals on the shores of Lindisfarne and trot along the top of stone walls in the Peak District. It's strange just how ubiquitous they seem to be, given their increasing rarity. An everywhere and nowhere bird. Perhaps it's because we so often remember these encounters – where minutes seem to stretch and centuries collapse – for the incredibly special experiences they are.

My favourite moment with a curlew is the first time I ever saw one here in the forest. I was fifteen and had just spent the night sleeping rough in the heather on Ragged Boys Hill. It had been a restless night. I'd been visiting some ancient yews, planning to sleep among them, but couldn't shake the feeling of being watched by something in the shadows. This growing sense of unease had made my skin crawl and eventually driven me out onto the heath, where a few hours later I was woken up in the pre-dawn gloom by the echo of an unplaceable sound. The heath was thick with mist but propping myself up to peer through the heather tops I saw the grey shade of a sickle-billed bird. I lay back down, listening to the curlew call out its territory from within the mist, catching the occasional fleeting shadow as it alighted. Sometimes a solid silhouette, at other times no more than an ink-washed blur. As the sun rose, I slipped back under the eaves of the wood and watched the bird use its long bill to preen its dew-soaked feathers before standing sentry again. I hadn't realised at the time, but there must have been a mate nearby. That was in 1990 and when I asked Andy whether that territory was still active, he told me they hadn't bred there for over a decade. The haunting call of the curlew is so evocative, it's almost predisposed to dwell only in our memories, but as I write these words, I realise how resigned and defeatist such an observation is. The curlew still has a place in our countryside and future.

By late afternoon, with the female back on the nest, the light is softening, the air has settled, and the sun has swung round, backlighting the grass. By 7 p.m. the landscape is glowing in beautiful evening light and the world is full of texture again. Grass stems heavy with seed glow like tiny silver spears and clouds of insects rise and fall in complex patterns only they understand. I decide to stay until sunset in the hope that the birds squeeze in a change of shift before dark.

They do and it's worth waiting for. The female lifts her head above the grass and the setting sun fills her eyes with gold. I hear the sound of the male approaching from the east and she leans forward, stretches her wings, and lifts vertically into the air, her white rump tinged with pink. Her head hangs low between powerful shoulders and her long legs trail behind her. She's a heavy bird and her back muscles work hard to provide the lift needed to clear the grass. Her long beak pivots like a weathervane in anticipation of the direction she wishes to turn, and I catch a glimpse of the distinctive snow-white V on her back as she gains height and drifts out of frame.

A few moments later, the male lands nearby and limps on to the nest. I stay until sunset, watching the huge red orb sink behind a ridge of tall conifers on the horizon. Like a mood change in a theatre, the moor is immediately bathed in soft purple light. I pack up as I wait for Matt to arrive. We walk out across the heath together while a curlew whoops and hollers behind us. The valley is theirs once more.

Thursday 4 June

The initial spike in visitors seems to have calmed down a bit, although Fritham car park is still packed. I step onto the badger trail that leads into the woods. Fallen blossom lies in a halo of white below the two hawthorns and I notice a cleared patch of leaf litter beneath the drooping branches of a beech. A scattering of fresh, gut-slick droppings shows where a solitary fallow doe has spent the past few hours lying up to chew the cud – probably pregnant lying low on her own, waiting to give birth.

The den site is almost entirely covered by bracken. By half past two in the afternoon the fox cubs are out, but I see only one at a time and can't tell whether it's the same one. There were three when I last visited six days ago and as the afternoon goes on my suspicions are confirmed: there's only one left. This is odd. There aren't any roads nearby, and they're not old enough to be ranging very far anyway. The only thing I can think is that they've either been killed by dogs or are now spending the daylight hours below ground. For my own peace of mind, I go with the latter.

A look through the lens shows me that the cub has grown a lot within the last week. Its weak grey eyes have hardened into amber to reveal their catlike vertical pupils. The russet coat is losing its woollen texture, the juvenile guard hairs are less noticeable, and the back of its ears are solid black. A pale white patch marks the centre of its breast and stiff black whiskers stick out prominently from the sides of a bright white muzzle.

It's amazing to watch the colour and texture of a cub's coat change in subtle rhythm with the transition of the seasons. The smoky-brown coat of a young cub blends in perfectly with the leaf litter and deadwood of early spring. As the bracken grows it casts strong shadows across the ground, and so the cub's markings become crisp, angular and more defined. It's as if the cubs come gradually into focus as they grow. A process that continues through summer as the sun strengthens and the cubs grow bigger beneath the rising bracken. The cub's coat will become sleek and its legs longer, growing increasingly gangly and teenager-like as the summer turns to autumn. By the time winter comes around, the male cubs will be searching for territories of their own and first-year vixens will be ready to mate. But for now, those distant, challenging days seem far off, and the remaining cub is doing its best to entertain itself in the absence of its brothers and sisters. Tugging at a piece of old bracken and ambushing imaginary prey in the leaf litter. It's clearly having trouble filling the hours and eventually sits down on its haunches to look around for something to do. Shrinking from the passing shadow of a buzzard, it seems a lot more cautious than before.

Looking in my direction, its ears fold back as it dips its muzzle to smell something on the breeze. It's up on its feet without a moment's hesitation and disappears beneath the soft-focus bracken. Tall foxgloves sway in its wake and I can tell it's walking steadily towards me. The cub has clearly smelled something and I'm pretty sure I know what it is.

Aware that this might be my final filming session with them, I brought the foxes a little present to say thanks. Sentimental, but I couldn't resist. Besides, it felt wrong to throw away the leftovers from last Sunday's roast. So, I brought the chicken carcass with me this morning and threw it out into the bracken to see what would happen. A fox cub munching on chicken isn't likely to make the edit, so I ignore the camera and enjoy watching for the sake of it. The bracken twitches as the cub comes closer in a zigzag pattern, homing in on the white carcass with remarkable accuracy. It grabs the food, trotting straight back into dense cover with head held high.

The roast had been seasoned with garam masala and any self-respecting urban fox would have tucked into it right there and then in the open, but it appears it isn't so easy to win the favour of a rural fox, and I don't see it again. The evening grows cold and I leave, not sure if I'll ever see these animals again, but grateful for the time I've been given with them. The rest of their story will go unfilmed, but I hope they do OK and that this post-lockdown world doesn't prove too hard on them.

Friday 5 June

We have decided to try to film the conflict between ground-nesting birds and crows, so Matt and I head out onto the wide-open heath south of Lyndhurst. There aren't any curlew here, but the valley holds a pair of lapwings, their black and white plumage flashing against the distant trees. It's still early, but rows of cars line the verges and forestry gates are blocked. Luckily, the one we need isn't, so I unlock it and drive through.

My phone rings just as I turn off the engine. Answering it, I happen to glance in the wing mirror to see a man striding down the gravel track towards me. He looks angry and is accompanied by a younger man, struggling to keep up. The older of the two seems to be coming in very hot, fuming at the ears like a *Beano* character. I'm still chatting and there's no time for me to get out of the van, so I hastily wind up my window. I have no idea who the approaching stranger is, or why he's so incensed, but it's clear I'm about to find out.

'Open the window!'

Motioning that I'm on the phone, I refuse while watching spots of spit land on the glass between us. He turns red and shouts into the glass again, 'I said, open the window. I ain't got no disease!'

Making my excuses, I hang up and try to placate him through the glass, but he's too far gone. He's hopping up and down inches from the window and the spots keep landing – a comical sight if it wasn't for the fact that we're in the middle of a pandemic.

'What are you doing here? It's closed. Open this window now, I said!'

Not bloody likely, I think, but then the penny drops: he must be referring to the campsite further down the track. It's been shut for weeks, but we probably look as if we're up to no good. As I redouble my efforts to explain, Matt gets out of his car. Rookie mistake. Mr Angry is on him in an instant.

'What's wrong with him?' he bellows, jerking a thumb in my direction. 'Some sort of chicken?'

Matt smiles nervously and backs away to get out of range. The bloke's not having any of it though and takes two steps forward to trap him against the side of his car. His face is now inches from Matt's, who I notice is keeping his mouth firmly closed while reaching into his car to grab our permissions.

The man takes a moment to read the permits and I see his hackles lower. With the wind taken out of his sails, he steps back, holds up his hands in grudging acceptance and marches off in a black cloud, disappointed at being denied a decent row. I raise my hand and his silent companion throws me an apologetic smile as he passes. Matt exhales and almost collapses against his car door in relief.

'Crikey,' I say, getting out of the van.

'Yep, crikey,' agrees Matt.

The forest is under a lot of pressure at the moment and the cracks are beginning to show.

Saturday 6 June

An impossibly huge moon adrift in the turquoise sky of very early morning. This is the last full moon of spring: the mead moon. Its name conjures up fertile images of honey and summer bounty. It flits through the trees alongside me before leaping into the clear sky as I emerge onto the heath. The open landscape lies bathed in milky-blue while a delicate pink glow turns orange in the east.

I open both windows to let the morning chill flow over me and fill the van. The smell of the forest perks me up. A lone birch shivers and heels to a stiffening breeze. It's bending the opposite way from usual, suggesting the wind has changed direction. Sure enough, as I slow down to take my turning, I feel a fresh northerly on my face. As I lurch through the potholes lying hidden on the corner, a roe deer leaps out of the heather to my left. He bounds across the track, straight-backed and stiff-necked, to land with perfect poise next to

the gorse. This is the resident buck. A dashing Dick Turpin-like character who often hangs out here at the crossroads. I hold up my hands
in a gesture of deference and wish him a good morning through the
open window. He turns his head and stalks away.

It's been several days since I've driven the next stretch of road
and since then a whole rash of passive-aggressive warning signs have
sprouted up along the fence line. It seems the Ghost of Forest Future
passed through here too: 'Residents Only'; 'Private Road'; 'Access
needed at all times'; 'DO NOT park outside our houses'. Other signs
are a little more direct, simply telling all and sundry to 'Go Away!'
An Englishman's home is his castle, but while I feel their frustration,
there's also something peculiarly unsettling and ironic about it all. A
thousand years on, it seems the forest is once again readying itself for
confrontation as locals dig in against the threat of invasion. Driving
round the next corner, I half expect to come face to face with a
shield wall. Not that I'm completely unsympathetic. The increase
in public pressure on the forest within the past two weeks has been
alarming. The keepers tell stories of impromptu raves and bonfires
in the woods. Abandoned tents, discarded nitrous-oxide capsules
and broken bottles. Glastonbury, Reading, Boomtown, they've all
been cancelled, but many would-be festivalgoers are taking matters
into their own hands and causing a great deal of trouble for those
responsible for controlling forest fires and keeping the forest safe.
Not only for people but for its livestock and wildlife also.

Something Andy once said to me comes to mind: 'I look at the
forest as a place that's been abused for centuries. Man has always
taken what he wants from it with little regard other than that.'

He was referring to the plundering of natural resources, but in
this day and age the emphasis is changing. We all need a release but
open-access areas like the forest are clearly under new pressures

that they haven't been expected to cope with before. Whether we exploit nature for financial, emotional or even spiritual gain, we are still exploiting it. And as our populations continue to grow, so do the levels of impact. Never one to shy away from the brutal truth, John Fowles summed it up perfectly in *The Tree*: 'We shall never fully understand nature (or ourselves), and certainly not respect it, until we dissociate the wild from the notion of usability – however innocent and harmless the use. For it is the general uselessness of so much of nature that lies at the root of our ancient hostility and indifference to it.'

As usual, entering the woods is like stepping through a portal. The silver trunks of birches glow and the bracken is now waist-high. Oaks reach out their arms to catch passing drifts of hawthorn blossom, while birdsong fills the air. Parking up in our usual spot, Matt and I begin our walk in. The breeze is cold and low, flowing over the contours of the ground. The nest is in shadow, but an adult drifts silently over to see who we are. There's no doubt we're recognised and, encouraged by the lack of alarm, I climb up quickly and quietly.

There's an echo of winter in today's weather. I peer out through the flap in my hide. These upper layers of the wood are vast and airy and the voice of a solitary thrush echoes through the vault like the whistle of a caretaker. A cow lows from the other side of the valley and the northerly breeze carries the muffled chest-thump of artillery from Salisbury Plain.

The wind continues to stiffen as the morning wears on. The top half of the larch is twisting in the wind, its branches gently rotating back and forth like the spokes of a wheel. The chicks are becoming active and one in particular spends a lot of time standing up, looking out into the wider wood. Neither parent is anywhere to be seen but, strangely enough, the chick's crop is bulging like a fluffy golf ball.

They must have either had an enormous meal last night or been fed a hearty breakfast just before I arrived. The nights are so short at the moment, the parents are probably burning the candle at both ends. The gloaming doesn't belong only to owls. Plenty of other raptors also hunt in the twilight world of dawn and dusk. The ultimate forest raptor, the harpy eagle, is particularly crepuscular. And what is a goshawk if not our very own mini-harpy?

In any case, the chicks seem perfectly content to sit and look around without keening, their white down giving way to the same grey of the orphans I filmed three weeks ago, their emerging pin feathers coming through too. The chick I placed in the surrogate nest will be fledging around now, but these three are several weeks away from that.

From the moment I saw that birch blowing on the heath this morning, I knew today might be sketchy. I'd hoped the weather would settle, but it has only grown worse and the reality of sitting in a tall, thin tree with 35 mph winds seething through the forest around me eventually grew too much. The northerly's continued to blow in unpredictable gusts, reaching a peak at midday when the tree lurched so badly I was knocked off my stool. The low, ominous hum of air approaching from the north had most impact, while the louder roar of wind curling round from behind, although sounding worse, was harmless – dispersing with a seething hiss like surf on shingle. The guy lines were a comfort – at least the tree didn't break – but they had the nasty habit of stopping the movement so abruptly that the erratic recoil was often worse than the initial gust. It's been cold, blowy and wet. Rickrack weather.

I'm on the ground now, having bailed out of the tree, in a hide hidden among the low-hanging branches of young hemlocks.

They're thrashing in the wind and my head is still spinning. The ground beneath me heaves on an invisible swell and my body feels as if I'm still 50 feet up. It's not ideal, but I have a clear shot up to the nest and it's better than nothing. The good news is that the male goshawk delivered food late morning and the chicks are still all safe. They came flapping and reeling towards him, though he quickly left, plummeting down to swoop away on a cushion of air. Where the female is I have no idea. Enjoying the feeling of being free again I suppose, making up for lost time. Hunting or bathing, or simply delighting in the feeling of being able to fly for the sake of it once more.

All seems quiet up there, although my low angle prevents me from seeing much. On the plus side, my new position is much less draughty, and my body has made the soft, dry needles beneath me deliciously warm. Any sensible animal would simply curl up and sleep out the storm.

Sunday 7 June

There's a swan huddled on the hard shoulder of the A36. It's almost exactly where I'd stopped to watch the muntjac feeding on leaves all those weeks ago, but now the traffic on this route home is fast and relentless. Only a fool would dodge the cars to rescue a bird, so I drive past, part of me hoping that the swan is already dead so that I don't feel compelled to do anything. But it isn't, merely stunned, and I can't ignore it.

Pulling into a lay-by, I turn around to go back and am relieved to see that the bird has recovered enough to walk. However, it is now waddling across two lanes of cars and trucks travelling at 70 mph. Brake lights flash and cars swerve, and it's only a matter of time.

Luckily, there's an open gate leading into a farmer's field directly opposite, so I put on my hazards and slow down to use my van to corral the bird towards the entrance. It works and I'm off the road before the next stream of HGVs whips past. Grabbing a blanket from the back seat, I jump out and usher the swan through the open gate before it can double back. It clearly wants to cross the road again and gets annoyed, flaring up to hiss with inflated neck and open wings. Preferring to be bitten by a swan than hit by a car, I push forward and eventually get it down into the field. It doesn't seem injured in any way and looking at a map on my phone, I see there's a lake on the other side of the road. The bird was probably on its way down to land when it collided with the overhead power lines or even a truck. To try to keep it in one piece, I run forward, swinging the blanket around my head. Having chased it as far away from the open gate as I can, I turn around to see a farmer watching me from his tractor in the distance.

Friday 12 June

I've spent the past couple of days resting and sitting out bad weather back in the forest. Heavy rain and deep rolling thunder: proper summer storms, the likes of which we don't see very often any more. A thick pewter-grey haze now hangs beneath dark, pregnant clouds and the air was so charged this morning that I had to go for a run. The heavens opened and I was drenched, but it felt fantastic: wild and invigorating. This is the first decent rain we've had for weeks, and what a release. I didn't realise how badly we needed it. The forest feels very different now. Softer, yielding, more reflective. Going to check on the nest ahead of filming there tomorrow, I noticed how saturated with smell the wood has become. The rain has released the heavy scent of moss and pine. Like adding a drop of spring water to a good Islay malt, the air is now thick with peaty goodness.

Saturday 13 June

The rain stopped around midnight, but three hours later the forest still feels muggy and subdued. Mist hangs in the trees and foxgloves nod under heavy pearls of water. Matt and I hear the female before we see her. She flushes the nest in a damp flurry of wet feathers and instantly dissolves into the watery half-light.

It's been exactly a week since my last session. The chicks are now a month old and, apart from their size, the most obvious difference is their patchy plumage. Pale down covers their heads and legs, but their backs are now prickled with emerging feathers. They still have a long way to go, but they are a little over halfway through their time in the nest and beginning to grow into themselves. They're also a lot more active. They can't yet fly but spend a lot of time kicking around, getting bored and fractious. An all-too-familiar sight for those of us in lockdown with kids of our own.

The nest itself is beginning to look tired and haggard. It's listing to the right, with the adults' landing platform noticeably flattened. A long fringe of shaggy twigs hangs down where a supporting branch has given way and the nest cup itself has long since disappeared. It needs a good shake-up before it gives up the ghost entirely and slides out of the tree. This is pretty standard. Goshawks are big birds and their nests inevitably sag over time. It's been in use 24/7 for over two months and like a threadbare mattress in a down-at-heel B&B, things are only set to get worse. From now on the pace of life in the nest will start to accelerate. Not only are the chicks more active, but the amount of food being delivered to the nest will

164

increase exponentially. Despite best attempts by the adults to keep their home clean, small scraps of meat will inevitably work their way down between the twigs and attract flies. Feet, gristle, fur: these less enticing titbits will get trodden in and rot. The fresh sprigs of fir brought in by the female are more important now than ever. Not just to help prevent the nest from disintegrating entirely, but for their sanitising effect.

When not jumping around and shaking down everywhere, the chicks while away hours by attacking springy larch twigs that bounce back when pecked. The chicks can't resist, jabbing and nibbling at the cones until – for the grand finale – they leap up to finish the job properly. A long leg darts out to seize a cone in a steely grip and the life is squeezed out of it with evident delight. Once satisfied it's dead, the chick glares at the remains as if challenging it to get back up, then either hobbles off to find another victim or lies down to rest. Cone killing is a tiring job.

Another favourite is trampolining, where a chick uses what spring remains in the knackered old mattress to bounce higher and higher while frantically flapping its unruly wings, knocking the other chicks off their feet. As in a pillow fight, the air is filled with wisps of down. The chicks' wings seem to have a life of their own and often flap without warning, dragging their owner across the nest. Clumsy yellow feet clutch desperately at twigs until, having safely anchored themselves, they can then really let rip. Occasionally the wings flap hard enough to pull them up to the point where it looks as though they'll soon take off. It's a bit like watching a rocket revving on the launchpad with its brakes still on. And on the few occasions they do let go, it's touch-and-go whether they land safely back on the nest at all. A plummet to the forest floor before they can fly would be the end for them.

By late afternoon, after numerous rounds of cone pestering and nest bouncing, all three chicks are flaked out and I'm beginning to wonder if their parents are going to appear at all. Then suddenly all three chicks perk up and start flapping their long, ragged wings in the limp fashion so characteristic of begging raptors. The crest of young feathers on their backs rises in excitement and they clamour to get towards the incoming food. Their mother lands on top of them, a dead pigeon in her right foot. The nearest chick immediately swamps her with outstretched wings. Judging from its feathers, it's the oldest, although not the largest. It quickly turns its back on its siblings, mantling the food to hide it. Shielding the prey like this has an out-of-sight-out-of-mind effect on raptors and sure enough the two siblings quickly lose interest and make no further attempts to wrestle it away. They stagger off to different corners of the nest, turn their backs on each other and sulk hungrily while the older chick eats its fill.

Due to a combination of asynchronous hatching and chick sex, there's usually great variance in size between raptor siblings and after getting a good look at all three of these guys, I'm fairly sure the oldest is male (slightly smaller but with more developed feathers). The middle chick (largest) is probably female, as is the youngest who is already the same size as her older brother. Although big, the two probable females currently have far fewer feathers. A result of more energy going into the growth of bone and muscle, I presume. Still, the only way to be totally sure of sex at this age is to weigh and measure the chicks – normally done as part of the annual ringing process.

Each year volunteers ring some 900,000 birds in Britain and Ireland as part of a coordinated research programme operated by the British Trust for Ornithology. Most of these are caught in mist nets as adults or fledged juveniles. By giving each bird a uniquely

166

identifiable leg ring, while also recording body weights and measurements, a demographic profile of population movements and size can be established. It also provides essential data on survival rates, lifespans, brood numbers and moult sequences via the recapture of previously ringed individuals. It's a programme that's been going in one form or another for over 110 years and has proved invaluable for conservation on both local and global levels. Birds know no political or national boundaries and the seasonal movements and migration patterns that ringing reveals are often astonishing and humbling.

Unlike smaller birds, goshawks and other raptors can't be caught very easily as adults. Not without considerable stress to both bird and person at any rate. So the accepted norm is to ring them as chicks in the nest once they are big enough to display sexual dimorphism, but not so old as to be able to fly away at the sight of an intruder. Or, indeed, attack said intruder. No one wants to be greeted by a faceful of talons as they poke their nose over the edge of an eyrie six storeys above ground.

About 65 per cent of all goshawk chicks raised in the forest since 2001 have been ringed. This is normally done as part of an annual tour of known nests in early June, but since goshawks in the forest have such a protracted and staggered season, there are inevitably some that don't happen because the chicks are either too old or too young. A visit to a goshawk nest can provide lots of other useful data as well: prey remains, nesting material and moulted feathers all have a story to tell. Unsurprisingly, though, Covid has put a stop to any goshawk ringing in the forest this year, so I am left guessing at the sex of our chicks.

So far, the smaller, but older and therefore nimbler male seems to be able to ensure first dibs by quickly intercepting food on arrival. By mantling and eating quickly, he can keep his younger sisters at bay.

But this will soon change. And as his sisters grow bigger, there will come a tipping point when he won't be able to hold on to the food any more. He'll be muscled out and slip swiftly down the pecking order. It's in his interest to make the most of it while he can.

Thirty minutes later, the female goshawk returns to the nest. She drags the half-eaten remains away from the male, who lounges back on his haunches with a full crop. Standing side by side, his two sisters then receive food from their mother in turn. By 7.30 p.m. the carcass is a limp red skeleton. The mother picks up the remaining scraps and flies the nest – a long pink foot dangling from her beak.

Sunday 14 June

An enormous blood-orange sun floats on a horizon veiled in mist. Ponies stand knee-deep in red heather and low-slung cobwebs dipped in gold wave lazily in the early-morning breeze. I'm in the south of the forest, back on the trail of curlew. I've been put in touch with Russell, a local conservation scientist who's been researching them here. He works closely with Andy and has solid knowledge of what's going on out there on the heath. He tells me that ravens have been mobbing the curlews we're going to see today in an attempt to flush out their newly hatched chicks. The parents have done well to protect their eggs through incubation, but now the hard work really begins and in the absence of large breeding colonies, this isolated pair has a big challenge ahead. Luckily, their young chicks are already highly mobile and have an innate fear of breaking cover, preferring to forage beneath the tall stems of mature heather and grass where their cloudy, mottled down helps them disappear. Nevertheless, ravens are almost primate-like in their ability to solve problems and I suspect they don't even need to see a curlew chick to make an accurate guess as to where it's situated – they just tune in to the behaviour of its parents.

The rising sun sets the heath ablaze as Matt and I walk out towards an ancient holly growing on its own. Nibbled by ponies and gnarled by time, it stands next to a low-lying pool, fringed with marsh. I've seen curlew here before, probing the wet mud for wriggling delicacies, and I'm hoping to catch a glimpse of their chicks should they return. As we arrive, it's clear there hasn't been any open water in the pool for days. The hot, dry weather has sucked it up into the sky and all that's left is an open plate of cracked mud and clay-fired pony tracks. My hopes aren't high, but I set up my hide regardless and Matt walks away to draw the birds' attention. I hear a curlew calling from the other side of the road that bisects the heath, but nothing comes close to the dried-up pool. An extensive marsh lies beyond the road and I suspect the birds are lying low in there, where the ground's still wet and the tall bog cotton shields the chicks from the sky.

The soft morning light has now hardened into a brittle glaze and it's like trying to film through the patterns of antique glass. There's also part of me that wants to give these birds as much room as possible. They're having a hard enough time as it is, and the success of our film doesn't rest on a few cryptic shots of their chicks. So, deciding to leave them well alone just in case they do wish to return to this side of the valley, I call in Matt and we pack up.

Just before we leave, I take a closer look at the old holly whose branches hang down around us. Its smooth skin is rippled with stretch marks and etched with the traces of pony teeth. Times get hard in winter, especially for ponies out on the heath, so they have to get creative in their foraging behaviour. Holly leaves are spiky for a reason, not that this will deter a desperate pony too much, but when heavy browsing pushes the leaves up beyond reach, ponies and deer often resort to nibbling the bark to get at the succulent tissues beneath. Holly bark also contains chlorophyll – hence its greenish tinge – and although it's harder work than eating normal leaves, there are times when the animals have little choice. Holly stems accumulate layers of damage and even now, as we near the peak of summer, the herring-bone patterns made by the gnawing teeth of deer as well as ponies are a sober reminder of how tough it gets between growing seasons. Walking round the tree to the other side, I find a pile of empty oyster shells neatly stacked in the crook of a stem. Obviously placed by human hands, they are an enticing mystery and I'm left to ponder their significance as we walk back to the vehicles.

By the time Matt and I are at the lay-by, there's a steady stream of cars, motorbikes and racing bikes on the road. Cursing myself for not buying shares in spandex before lockdown, I pull out to crawl up the hill behind a group of twenty cyclists. In an attempt to rescue the day, Matt and I have decided to recce some future locations, so we're heading north to Fritham. But the traffic is so thick, and the roads so choked, that we think again. Turning off onto a back road, we go instead to collect a trail camera we left near our original fox earth back on that cold, rainy day in April. The idea had been to monitor the earth ahead of filming it, but we forgot all about it once we started filming foxes elsewhere. It's been clicking away ever since. Day in, day out for six weeks. So it'll be interesting to see what it's captured, if anything.

Trail cams are extremely useful tools, especially when dealing with elusive animals or sporadic events, and are particularly good for capturing images of animals like big cats that range over vast territories. Closer to home, I use them a lot to help build a picture of animal movements. They can also be a lot of fun. Like the pro-verbial box of chocolates, you never really know what you're going to get. And although Matt's camera certainly has a treat in store for us, it isn't what we're expecting.

We press play and watch as a mysterious, low-slung animal weaves its way towards camera. The long-pointed snout, pale chest and luxurious dark coat leaves us in no doubt: it's a pine marten. And a particularly handsome one at that. Given that these enigmatic creatures have been absent from the forest for well over a hundred years, it's not what either of us was expecting to see, so leaning in a little closer, we scrutinise the footage in rapt silence.

Moving with purpose along a fallen log, it's following a scent trail. Sniffing its way towards camera before dropping down out of frame. We replay the clip to make sure our eyes aren't deceiving us. It's a beautiful animal. Its black twinkling eyes reveal what feels like a playful intelligence, its powerful shoulders and thick forearms speaking of an active arboreal life. I can clearly see the strong claws that enable it to move up trees just as fast as it can run down them. Its blunt ears are broad and trimmed with lighter fur, while its long bushy tail is surprisingly fox-like.

The overall impression is of a large, stealthy weasel, which is hardly surprising given that they are members of the same family, along with stoats, polecats and otters. The sinuous, fluid body profile of these five native mustelids are so characteristic of their clan that it makes you wonder where on earth its sixth member, the badger, came from. Lovely as badgers are, not even the most ardent fan could ever call Old Brock stealthy and sinuous.

The marten reappears a few clips later, this time at night. Jumping up onto the log from the left, it crouches as if on the hunt, then leaps back down into the shadows. Viewed side-on, its long body has the appearance of two halves, as if its articulated rump has a life of its own and is about to wander off in a different direction.

This is my first glimpse of a pine marten in the forest, but not Matt's. He took a rare photograph as one crossed an open glade last year, but that was several miles south from here and a comparison of throat markings reveals they are two different animals. Coupled with a recent surge in these animals being killed on roads and a smattering of footage from other trail cams, the evidence suggests that the forest is in the process of being recolonised after an absence of over a century.

In fact, pine martens were pretty much extinct all over England by 1900, the usual story of persecution and overhunting being mostly to blame. Pine martens have an even more eclectic diet than goshawks and are certainly not above pinching eggs or grabbing game birds – a habit that didn't exactly endear them to gamekeepers and country-sport-loving Victorians. Even prior to this, martens had always been actively hunted, trapped and snared for their thick, soft pelts – particularly during winter when their pelage is especially long and silky. It's almost certain that many of the medieval nobles who hunted Nova Foresta a thousand years ago wore cloaks trimmed with their fur. Marten pelts were valuable enough to use as currency and in 1086 the Domesday Book records the city of Chester paying its tax bill with 150 of them.

I love the fact that martens are making their way back into the forest, but the obvious question is: where are they coming from? There have been licensed reintroductions and restorations in Wales, Shropshire and the Forest of Dean, but the New Forest is so far removed from these – and a world away from the main stronghold up in the Highlands of Scotland – that it seems to suggest deliberate unauthorised release.

Martin Noble has been studying their resurgence. The first sighting in the New Forest was as far back as 1993, but it's really within the past five years that they've managed to gain a tentative foothold. In a 2018 interview with local news, Martin mentioned that a recent roadkill animal was found to be of Czechoslovakian origin when DNA-tested. In his words: 'I can't imagine that it's come from anywhere except a private collection somewhere in this country.'

Both licensed and unlicensed releases are clearly under way and, as with the eagle owl, I can't help thinking that yet again there are strong parallels between the return of the marten and the goshawk. This aside, goshawks and martens are certainly united on one thing: their love of squirrels. In fact, although preferring to ambush when they can, martens are one of the very few arboreal mammals capable of running squirrels down in a treetop chase if necessary.

Squirrels are world-class acrobats and nothing short of a revelation when they really get going. As someone who spends a lot of his life filming in trees, I have nothing but admiration for them. Time spent watching them foraging or at play is always a joy. They make it all look so easy, as virtuosos generally do. The forest canopy is probably the most complex environment on earth, so any predator wishing to hunt squirrels has to be able to create trapping opportunities and to dodge, swerve, tuck and roll on instinct. It also has to be able to calculate shifting parallax at blistering speed, make split-second life-or-death decisions and have an unswerving determination to stay on the chase.

Just like the goshawk, the marten has evolved to deal with such challenges perfectly – albeit in a very different way. Everything about its body is designed for exploring cavities and weaving through branches at high speed: a low centre of gravity, semi-retractable claws (the only mustelid with this catlike feature), binocular vision to judge distances accurately, and ankle joints that swivel through 180 degrees enable it to climb downwards head first.

I once watched two yellow-throated martens foraging in the high tropical canopy of Borneo. Their freedom of movement was unparalleled. I'd never seen anything like it. I was there to film gibbons and although the apes were masterful, these two mustelids took tree climbing to the next level, twisting, looping and spiralling through the canopy like otters in water.

There's a hypothesis in evolutionary biology called the Red Queen effect, used to describe the ongoing 'arms race' that exists between predator and prey as they constantly evolve to try to outpace each other. But where do you go when you're a squirrel and your own personal predator is something as awesome as a marten? Well, in Borneo you evolve the ability to glide, which is how flying squirrels avoid becoming a meal, although I don't think our own greys have quite worked this one out yet.

The last time I saw Andy, he showed me a trail-cam shot of a British pine marten ambushing a grey squirrel. Having climbed a tree, the hunter pauses outside a bird box to listen. Once it knows the current owner is home, it flips open the lid and plunges in head first to grab what's inside. Its hindquarters stay outside the box and, judging from the desperate scrabbling going on inside, the marten doesn't have it all its own way (squirrels can't half bite). But when prey is cornered like this the contest is a foregone conclusion and sure enough, after a short, sharp struggle, the marten re-emerges, dragging a limp grey body.

It's a graphic example of just how determined and resourceful a predator the pine marten is and although one less squirrel to wreak havoc in a forestry plantation may be a good thing, it's a fact that most successful predators are opportunists and rarely stick to a set menu of so-called 'pests'. In fact, the interesting thing about martens is that they're one of the few predators that are agile, swift and bold enough to raid a goshawk's nest.

'Oh, they'll kill the chicks,' Andy told me, 'no doubt about that. Once those birds have got three-quarters-grown young in the nest, and both birds are away hunting, then the pine marten would be up there and have a chick out the nest as quick as a flash.'

On the other hand, a female goshawk might make short work of a marten if it could catch one unawares. Again, it's all about opportunity. Ecosystem relationships are highly complex. Food-chain hierarchies and preferences are often fluid, and every gain or loss has a knock-on effect.

'These predators were here years ago,' Andy explained, 'and it's probably only right they should be back. But you have to accept that when you bring these things back – just like goshawks themselves – it will have an impact. But how do you know what's the norm? Back in the days when we were planting up the forest and there were hundreds of acres of new plantations, we had a lot of woodcock here. But when those trees grew up, they became less suitable for sheltering woodcock and other species come in. Some things decline and some things do better. So, you're on this continual line where things are changing all the time and when you chuck a new species back into the mix it's obvious that some others are going to suffer in the short term until that balance becomes settled again. It will settle, but it takes time and needs to be carefully monitored.'

If nothing else, what the recent reappearance of both goshawk and marten has perhaps shown is that these things sometimes have a momentum of their own. There are clearly people out there who want to see certain native species back in the countryside and don't want to wait for official blessing.

A quick look at a map of the New Forest reveals the echoes of various other species lost over time – Cranes Moor being a classic example. But perhaps the most famous, and certainly the most poignant, is the Eagle Oak, an ancient tree located on Vinny Ridge in the heart of the forest. Although the naturalist William Gilpin wrote of a pair of golden eagles nesting in the area in the 1700s, the Eagle Oak actually takes its name from a sea eagle shot from its branches by a forest keeper in 1810. By the time Gerald Lascelles was appointed Deputy Surveyor of the New Forest seventy years later, it seems that with only 'four responsible keepers left', unlicensed shooting was rife among the unruly 'under keepers'.

Following in the wake of the last sea eagle, it seems other raptors had continued to suffer from an almost systemic culture of persecution. In Lascelles' words: 'Everything in the shape of rare birds that they could get hold of they regarded as perquisites. With some trouble I discovered the Southampton bird-stuffer who was in the habit of regularly paying them 3s. 6d. per head for all kingfishers he could get. Everything in the shape of a bird of prey was, of course, looked upon as vermin, killed, and if possible sold.'

Although 'everything was in a state of chaos', Lascelles quickly 'set to work to clear out what was verily an Augean Stable'. While the honey buzzard was given a last-minute reprieve and saved from complete eradication, it seems that the wild goshawk was long gone from the forest by then and it was certainly too late for the sea eagle.

The year 2019 saw the highly publicised, not to mention highly welcomed, return of this species to the Isle of Wight, barely an eagle's glide from the forest. A reintroduction programme, initiated and delivered by the Roy Dennis Foundation in partnership with Forestry England, saw initially six birds released. Natural England has granted a licence for the release of up to sixty juveniles within a five-year period, and it's hoped that this will provide the basis for a viable breeding population. The Isle of Wight was chosen due to its historical association with the birds – the last known breeding pair nested on Culver Cliff in 1780 – but also because of its strategic position in terms of linking existing populations in Scotland and Ireland with those on mainland Europe. For example, Denmark, barely a day trip away for such a long-range flyer, currently has around 100 breeding pairs.

As the Roy Dennis Foundation clearly states, this is a species that was systematically removed from the ecosystem by us through sustained and relentless persecution from the Middle Ages onwards, and since 'many parts of southern England remain highly suitable for the species', it seems only right that we should make amends: 'This project provides an opportunity to restore a population of white-tailed eagles to parts of its former range in southern England from which it was eradicated entirely due to the influence of man.'

With a wingspan of up to 2.5 metres and a weight of up to 6 kilograms, a female sea eagle is approximately three times the size of a female goshawk. The largest species of eagle in Europe, in fact. And the fact that such a massive, high-profile raptor has been so warmly welcomed back by the vast majority of people on the island and neighbouring mainland counties seems to indicate how popular such reintroduction programmes can be. Of the 10 per cent of people on the island who didn't want to see them back, the main objection was due to its unknown impact on livestock, particularly sheep. So far, no lambs have been taken and why should they be? The sea eagle is a generalist raptor, switching easily from the hunting of water birds and fish to the scavenging of carrion, depending on seasonal availability. Given that the coastal regions fringing the Solent are some of the richest in wildfowl and fish in the whole of the UK, livestock (and pets, as one unhelpful newspaper article suggested) should remain safely off the menu.

For me, one of the most intriguing, not to mention exciting, prospects of the whole endeavour is that we might one day have a pair of eagles back in the New Forest. Despite their famous association with Culver Cliffs, sea eagles often nest in large trees through much of their range. Since Forestry England is a partner in the initiative, I can't resist asking Andy his thoughts on whether they might return to breed on the mainland: 'We've seen white-tailed eagle in the Forest and we know from the satellite data that they've been regularly tracking across here, but they range across such a distance, the impact they're going to have is going to be negligible for at least a while, I'd have thought. But I think there's a good chance that they could return to the forest. If I saw it in my lifetime, I'd be very happy.'

Which is good enough for me. I remember seeing my first white-tailed eagle soaring high above the silver-washed sea off the west coast of the Isle of Skye. It was a scorching hot day and the bird was nothing more than a silhouette against the sun. I'd seen golden eagle earlier that morning, but that hadn't prepared me for the sheer size and bulk of a sea eagle. It owned the sky, there's no other way of putting it, and the mere thought of them one day deciding to breed in the forest is enough to give me goosebumps.

Having collected the trail cam with its unanticipated visitor, Matt and I decide to relocate it at a large wood ant nest, in preparation for filming there later in the summer. When I found it, there were signs of the nest having been recently visited by a green woodpecker. Known locally as the woodnacker, these birds spread themselves flat on a heaving nest to let the ants crawl all over their splayed-out wings. The ants remove parasites and sterilise feathers with their formic acid, before becoming a tasty snack for the woodnacker. A great bit of behaviour to capture, but another scenario where the use of a trail cam could save days of waiting.

It's a few miles away in a different part of the forest but the road south is totally choked. By the time I realise my mistake it's too late and I'm part of the problem. Boldrewood is a beautiful area with lovely walks; it's always been popular, but the roads here are narrow and hemmed in by potholes. A train of parked cars now lines the verges, narrowing the lanes yet further, and the frustration of drivers being squeezed through such a tight bottleneck is clear. There's a queue to get through the cattle grid and drivers vent their anger by slamming up through the gears when they finally break through, despite the children trying to cross the road from between parked cars. A motorcyclist takes a short cut through the families picnicking on the grass and one driver loses his temper completely, sitting belligerently in the middle of the road while his poor partner pretends to ignore the angry horns around them.

Monday 15 June

It's now a month since the easing of lockdown began. Following the road onto open heath, I enter heavy fog. It's 3.45 a.m. Thick forest looms on both sides and for a few seconds I'm accompanied by a swift-flying bat as it zigzags through the gap between road and trees. The leaves of whitebeams and willow flash pale as I pass and, entering Lyndhurst, I catch the tail of a fox as it crosses the road in the same place the other was killed two weeks ago.

Whitemoor Glade is full of mist, moths flutter in my headlights and the smell of the forest fills my car: earth, leaf mould and pony dung. An evocative odour that I find strangely comforting.

At Highland Water, a white fallow doe stands glowing in the blue half-light. She's on her own and turns to look at me as I stop and open the window. She's heavily pregnant but shows no sign of alarm so I turn off my engine and watch her in the pale twilight for as long as she'll let me. Finally, she licks her nose, glances off into the trees and turns her head to leave. As silent as a ghost and just as ethereal. There's a good chance her fawn will be white too.

The open heath is so layered in fog that it's hard to see where land stops and sky begins. Five fallow bucks in velvet stand like statues among the dew-covered gorse, but there's no sign of Dick Turpin this morning. The solitary birch stands silhouetted and dog rose flashes pink in the hedges. The mist rolls up against the eaves of the wood but goes no further.

Leaving the heather fog behind, the string of 'Private Access' and 'Stay at Home' signs seem to be multiplying and now there are also piles of wood placed to prevent cars from stopping. It's beginning to feel a little like Royston Vasey down here.

Passing through the portal, goshawk wood is muted, soft and welcoming. Bird song in the air. The melodies are relaxed and still beautiful, but the dawn chorus lacks the Vivaldi exuberance of six weeks ago. The giddy rush of spring has mellowed. The goshawk chicks will soon be fledging, and another breeding season will be over.

It's 9 a.m. and so far it's been a good morning. Both adults visited the chicks to deliver small anonymous snacks. It's a still day with hardly a breath of wind so I linger at the tight end of the long lens. It feels good to film some close-ups of the chicks. I've just lined up and focused on the biggest when she ducks down, popping back up into frame with the tiny head of a robin impaled on the tip of her hooked bill. Nothing else, just its head, cleanly severed as if by scissors. It's a chick; its puffy red eyes are closed as if sleeping. The goshawk – suddenly enormous, cold and strangely mechanical – grips it by the end of its lifeless beak. The robin's pimpled flesh is covered in grey wisps of down and I can clearly see the delicate flap of an ear behind its bruised eye.

The robin was no more than a week old when the goshawk paid its nest a visit. It's a pitiful sight made all the more poignant from knowing that the chick would have instinctively reached up to beg for food as the hawk's shadow fell across it. This is the goshawk way. For a goshawk, it sometimes seems as if life is simply nature's way of keeping meat fresh, and I'm not in the mood for it today. The endless stream of corpses arriving on the nest feels repugnant. I film the juvenile choking and gagging on its grisly meal. It's regurgitated twice, before finally staying down. The rest of the robin's soft blue body lies to one side, attracting flies. It's sometimes hard to see the beauty in nature.

Tuesday 16 June

I'm amazed to discover an active buzzard nest 300 metres away from our goshawks. They too are in a larch, but theirs stands within a beech wood further down the ridge, separated from the goshawks' dark spires by an open patch of clear fell. I remember Andy telling me that this ridgeline had always been popular with raptors and that the original nest, on which the goshawks built theirs, was previously used by buzzards for years beyond count. When the hawks muscled in, the buzzards vacated. I'd assumed they'd left completely but am wrong. It's a good opportunity to grab a few shots, so I am now sitting in a ground hide waiting for an adult buzzard to deliver food. It's a large nest, almost as big as the goshawks', but far more exposed and easier to spot despite the fact that it's managed to avoid our detection these past two months.

Buzzards are now our most common, widespread raptor. More than 70,000 pairs breed in Britain compared to the goshawks' meagre 400. That's almost 200 times as many, so it's no wonder that today buzzards seem almost ubiquitous. They're certainly a welcome sight in the valley where I live and a summer's day in Somerset doesn't feel complete without them wheeling on the thermals. I love both goshawks and buzzards, but other than their general body size they have precious little in common. In fact, the closer you look, the more profound their differences become.

Keywords in the goshawk lexicon include wildness, intelligence, focus, ferocity, secrecy, stealth and power. Buzzards on the other hand scavenge for worms. OK, that's a bit harsh. Buzzards are also successful hunters but, when all's said and done, as avian predators go, the buzzard simply isn't in the same league. Think dog vs wolf, biplane vs Spitfire or breadknife vs Samurai sword. On the flip side though, their high numbers clearly indicate that buzzards are doing something right despite not being the sharpest tool in the raptor box. Pretty much anything is on the menu for a buzzard. Worms in winter, yes, but everything else from rabbits to grass snakes are also regularly taken in season, including the eggs and chicks of other birds of prey. Which is why the incubating goshawk seemed to shrink a little and gaze skyward every time a buzzard slid over the canopy high above. In short, they are extremely successful generalists and, like all raptors, they know how to take full advantage of any given opportunity.

Despite their familiarity though, buzzards can be surprisingly tricky to film at the nest. Goshawks are hard, but once they are locked into their breeding, they generally take well to the filming routine. Buzzards remain twitchy throughout and I'm reminded of this when the first adult returns. It's the female: huge, soft and beautiful, her plumage a subtle blend of chocolate and cream. She perches motionless on the edge of the nest, peering down at my hide. She's not alarmed, just curious. The worse thing I could do now is jog the camera lens. Any movement at this point would snap the tenuous thread, so I remain perfectly still and quiet until her attention turns to her large solitary chick. A few minutes later her mate arrives with a mouse dangling from his beak. He alights on a branch next to the nest, giving me a clear view of his blunt, rounded tail and bright yellow legs: skinny, short and feeble compared with a goshawk's. He passes the mouse to the female, who takes it gently in her own beak, the tiny corpse dangling by a front paw. She barely has time to offer it up before her full-grown chick snatches it and down it goes in one. The male leaves and the female soon follows. The chick is left alone to wobble and teeter on the edge of the nest, making the same staccato vertical jumps as the three goshawk chicks nearby.

Friday 19 June

Lloyd Buck wears a dark-blue boilersuit and sits on a plastic chair inside a small aviary. We're chatting about the pandemic, but I'm distracted by the starling perched on his head. Its name is Jack and Lloyd's training him up: offering little titbits as the tiny bird pecks affectionately at his ear. The boilersuit isn't essential, but judging from the white splashes on Lloyd's shoulder it's a prudent move. We continue chatting, setting the world to rights: increased pay for NHS staff; lack of faith in politicians, the usual stuff. Eventually, however, the conversation comes round to goshawks: 'So, I'm not a falconer,' he says. 'We're more like bird specialists and our approach to birds is very different, but what I would say is that there's a wildness in goshawks. There's something in them that's different from other raptors. You can't stop that. You see it in them.'

Despite not being a falconer in the strict sense of the word, there's not a lot that Lloyd and his partner Rose don't know about raptors. In fact, there's not a lot they don't know about birds in general, having spent the past twenty-five years hand-rearing and training falcons, hawks, owls, eagles, corvids, waterfowl and songbirds to fly alongside camera. Every time you see David Attenborough alongside a skein of geese skimming the surface of a Scottish loch, or a peregrine stooping alongside a skydiver freefalling from a hot-air balloon: that's Lloyd and Rose Buck.

Such memorable results require boundless patience and dedication, not to mention love and commitment. A golden eagle can live for forty years in captivity, a raven for twenty. So, each bird inevitably becomes part of the family. Such close relationships result in an intimate understanding of what makes their birds tick. Not just as a species, but as an individual. And it's this rare insight into the minds and lives of birds – of goshawks in particular – that I'm keen to hear more about.

Rose and Lloyd have recently taken on a newly hatched gos chick. Lottie hatched on 11 May, four days before the first chick on our nest, and I'm hoping to grab a few shots of her today. They also have a wonderful one-year-old called Mabel and, prior to this, raised another female called Ellie. Ellie appeared in numerous films. She was a regular on BBC's *Springwatch*, wowing audiences with her uncanny ability to fly through the smallest of gaps between branches. I'd filmed Ellie myself shortly before she died peacefully in her sleep at the grand old age of sixteen. Goshawks are very much central to the Bucks' way of life, so while Rose readies Lottie for the camera, I ask Lloyd whether he thinks they have a favourite prey. He tells me they're guaranteed to have preferences of what they like to catch and how they like to approach a hunt: 'They'll have a way that they've perfected, and their individual style will come through. Different birds will be better at some things than others, because their body shape, size and weight will give them an edge.' Some use tight gaps and surprise attack, while taller, leaner birds might prefer to fly things down over distance.

'I personally believe that a lot of their decision-making is done before they even take off. So, if they're sitting somewhere and an opportunity presents itself and they know they can slip out unseen, they'll already know where their best route is to stay hidden until the last second. They'll know where there's a gap they can cut through; they'll know where there's a hedge they can pop over, and trees they can go around. They'll know the ideal ambush point and would have already completed the hunt in their mind before they even begin.'

Like a racing driver visualising their perfect circuit, it seems the goshawk might play out its high-speed chase before even opening its wings.

Eric Ashby once told me he believed there were two basic types of foxes: fur and feather, meaning those that specialised in hunting birds and those that focused on mammals, such as rabbits. It's an intriguing theory, so I ask Lloyd whether he feels goshawks could be similar, or whether they're just very opportunistic. He thinks it can cut both ways, that they can tackle a huge range of prey, 'but if they can be selective, then I think they will be. Outside of the breeding season, I'll guarantee they're very particular.'

A few minutes later I'm looking down the camera lens into Lottie's pupils, black and fathomless. A movement catches her eye and my reflection slides across the sky as she turns to look at Rose. It's clear that Lottie is Rose's bird and that the imprinted five-week-old is infatuated with her surrogate mum. She returns to stare into the lens. One pupil slowly dilates more than the other, then both snap back into line as she cocks her head and peers up at a buzzard almost beyond sight above us. I spend the next hour filming her with Rose's help as Lloyd concentrates on Jack, but there's still one important question I want to ask before I leave, although I suspect I know the answer already:

'How easy is it to accidentally lose a trained goshawk?'

'Very!' laughs Lloyd. 'Not as easy to lose as a sparrowhawk, but it's very easily done, especially if you've got a bird that's hunting wild prey. If they make a kill and feed themselves up and stay out overnight, you're in trouble. The chances of getting that bird back go down massively, especially if they make another kill the next day. At that point they realise they don't need you and they're gone. Once they've got their hunting eye in, they just don't need humans any more. Most falconers really know what they're doing, especially goshawk owners, but even the best person with the best experience can still lose a gos. Especially in a hunting scenario.'

Words that echo Gerald Lascelles' own about his favourite goshawk Shelagh over a century ago: 'It is, alas! so easy to lose a good hawk; and the better she is the harder it is to recover her, for a good hawk is never hungry.'

With the filming finished, Rose picks Lottie up carefully, swaddling the fragile chick in her arms. What I'm witnessing here is the start of an intense and intimate lifelong relationship between two humans and one bird. Everything I've ever learnt about goshawks comes from my own experiences with them in the wild, as an outsider looking in. Lloyd and Rose's perspective couldn't be more different. Their close working relationship with these birds has given them privileged access to a depth of understanding simply impossible for me to achieve. In many ways they are experiencing the goshawk's world first-hand, from the inside out.

Reversing my van back up the grass track towards the main road, Lloyd leans on the access gate and motions for me to wind down the window: 'What I would say is there's no other bird of prey on the planet that's as successful and versatile as a goshawk. There is no other raptor that's as good at what they do. Their kill ratio is really high. There's nothing else comes close.'

In hindsight, the rise of a naturalised goshawk population seems almost inevitable. A fiercely independent, opportunistic, intelligent fugitive perched on the threshold of a new world ready for the taking. For a bird like this, the well-stocked virgin woods of the New Forest must have presented almost limitless possibilities. It's hard not to think that destiny – a cascade of unintended consequences through the centuries – had a hand in the whole thing.

Monday 22 June

The summer solstice was on Saturday and it barely grew dark last night. Rather, the sky sank into a deep pearlescence before growing lighter again four hours later. My pre-dawn drive across the forest is sublime. Bratley Plain lies beneath a moon mirage of golden mist and dew-soaked cobwebs hang heavy in the heather.

I'm settled in the hide by 5 a.m. and the sun rises ten minutes later. Gone are the sultry red rays of April. The sun now leaps into the sky to scatter light through trees like a prism. The two female chicks lie dozing on the nest, but there's no sign of their older brother. I'm worried he's gone for good. It's been a week since I was last here and a lot can happen on a goshawk nest in that time. But an hour later he comes careening in from a neighbouring tree, where he's clearly spent the night. Ragged wings wobbling, body swaying, long legs dangling, he collides heavily with a branch below the nest, pitches forward as if to swing upside-down, then regains balance. Recovering composure, he nonchalantly fluffs up his patchy plumage, dislodging wisps of down that float away through the still air. He's the first of the siblings to have left the nest. At this stage, he's using a mixture of hops, glides and barely controlled tumbles to move between branches, but he's taken his first steps towards independence and his skills will only improve. This is the 'branching' stage, a transitional period that all raptor chicks go through as they start testing the bonds that tie them to the nest. Their wings aren't developed enough to provide sustained lift, so they gain height by hopping from branch to branch, 'laddering' their way up to a height from which they can use gravity to glide down to another branch or neighbouring tree. It involves a lot of wobbling, a lot of courage and an awful lot of crash landings. Nevertheless, the whole process is a self-fulfilling prophecy since every attempt helps strengthen wings and build flight muscles. It's only a matter of time until powered flight is achieved and from then on independence gathers pace.

I'm not surprised that the male is the first to take the leap. Being the oldest, he has the more developed feathers – and why take the risk of hanging around two enormous, perpetually ravenous sisters when you can maintain your distance and simply return to the nest when dinner is served? The two females are still several days behind him, but as the morning wears on they too start laddering up and down between branches. They haven't made the leap to another tree yet, but it won't be long before they do.

The process of fledging, of leaving the nest for good, is often a drawn-out, protracted affair. Unlike many songbirds, which simply pitch over the side and continue their development in thick ground cover, raptors continue to return to the nest to eat food delivered by their parents. Their independence is linked to their own hunting abilities, but for the first few weeks they come and go and the nest remains a focus of attention for a good while after they've effectively left home. A bit like a student popping back for Sunday lunch or to use the washing machine. It takes a long time for a raptor to learn how to hunt properly and although goshawks have a better kill rate than many, it's still a steep learning curve. The parents themselves wield great influence by rationing the food they deliver. As any falconer will tell you, hunger is a great motivator and adult raptors often use this to catalyse the fledging process. The adults don't want their chicks to starve, but neither do they want them lounging around in the nest expecting to be fed for the next few months. By gradually reining in the food, they nurture the chick's own instincts. It's a fine balance that takes some of the larger raptors, such as harpies and the Philippine eagle, months to accomplish successfully. It's essentially the same process used by austringers when training a hawk to fly to the fist. The ability to control food is a primeval power that can be used to forge or break immensely strong bonds of dependence.

All three chicks are now fully covered in brown juvenile plumage. Scraps of down still cling to their heads like dandelion seeds and white wisps poke out from beneath breast feathers, but for the most part they're well dressed. Not that they look anything like their parents. In fact, juvenile goshawks look so different during their first year that the 'father of modern taxonomy', Carl Linnaeus, even assigned them a different Latin name. I'm sure that a quick conversation with any austringer would have enlightened him, but there's no denying that juvenile northern goshawks look extraordinarily different from the adult birds. The iconic wintry-grey plumage doesn't arrive until they are a year old and undergo their first moult. Until then they remain cloaked in an autumnal blend of copper and bronze. Warm hues waiting to be tempered into cold steel. It's not just the colours that set them apart, it's the way they're shaded. Whereas an adult's breast is etched with finely layered horizontal bands, the offspring are dabbed with vertical teardrops, which to my eyes makes them look more like Gerard Manley Hopkins' famous 'dapple-dawn-drawn Falcon' than a hawk. Even their eyes won't change from grey to yellow until a couple of months after leaving the nest.

There are many theories why the juveniles look so different during their first year. Most relate to a need to engender tolerance among the fiercely territorial adult birds they encounter while dispersing in search of their own place to live. There is some evidence to suggest that adult goshawks may be less likely to challenge another bird passing through if it is a sexually immature juvenile without a territory of its own.

All this lies ahead for the three goshawk chicks, but for now all they have to do is wait for their flight feathers to grow long enough to provide that elusive sustained lift. In the meantime, there's a whole network of branches to explore.

By late morning the sun is hot on the nest and the two sisters lie low. Their brother perches gargoyle-like on the branch below, but none of them is moving and from a distance the nest looks devoid of life. I'm just watching a goldcrest hang upside-down to pick insects from a cone when the female goshawk explodes out of nowhere, trailing the limp body of a large, freshly killed squirrel. Folding her wings to slide seamlessly through a gap in the branches, she spills the air and practically throws the dead animal at the two chicks who instantly rouse themselves into a chaotic melee of flapping wings. Their hackles are up, and they clamber and flounder over each other in their eagerness to get at the food. Their mother staggers back and is shouldered off the nest by the larger chick, who accidentally grabs a fist full of twigs in the confusion. Just then, the male chick piles back onto the nest, barrelling into his sisters and knocking them off their feet before deftly throwing his wings over the squirrel like a cloak. The females lurch back to their feet, not quite sure what hit them, but it's too late, their older brother has completed a masterful *coup d'état*. The females stand to one side pretending not to be interested while the smaller male glares with crest-raised defiance. Turning his back, he collapses on top of the carcass to catch his breath. One female hops onto an adjacent branch and the other begins to preen. The male stands up, dragging the squirrel off to one side before pinning it down to tug at its throat. I catch sight of the rodent's long yellow incisors, protruding from a bloodied mouth, and all hell breaks loose. This time it's the adult male bringing in a thrush, which is snatched from his talons by his larger daughter before he's even had time to land. The other female makes a lunge for it too, but her bigger sister has a firm grip and hops away with the tiny corpse dangling from her beak like a rag doll. Holding out her huge drooping wings like a phantom, she tugs at the thrush and doesn't stop until every last scrap is gone. There's still plenty of squirrel to go round and the other sister moves in to take over once her brother clambers off.

By 5 p.m. the chicks have retired to their chosen branches. With bulging crops and drooping eyelids, they waggle their tails contentedly while shaking out their feathers. Then, one by one, they raise a bloodied yellow foot to perch on one leg and fall asleep.

Tuesday 23 June

Dull, plodding thoughts behind dry, stinging eyes. Scattered shards of early sunlight rotating across tree trunks. The two female chicks stand on the nest dozing but once again there's no sign of the male. A woodpecker is calling from behind me, but the woods are full of other, more alien, noises this morning. Metallic clangs and groans of distant forestry machinery accompanied by the staggered footfall of MOD shells exploding on Salisbury Plain. An H. G. Wells soundtrack, conjuring up images of towering machines and chaotic dystopia. A strange start to the day and a world away from lockdown.

The two female chicks are now awake and preening. The air around them is full of white mealy dust and gossamer down. The larger starts a series of running hops from one end of the nest to the other. She's all legs and wings as she gallops to and fro, flattening the twigs yet further. The nest is looking pretty shabby now. Deflated and listless. After two months of 3 a.m. starts, I think I know how it feels.

Wednesday 24 June

I'm sitting at the dining table of my rented cottage. The front door is wide open, and I can hear the squabble of starlings and sparrows as they feed their young in the eaves of Matt's house on the other side of the fence. The sun is shining, and I'm watching a robin pick through the twigs beneath the hedge. I'm on my third cup of tea and my map of the New Forest is laid out on the dining table in front of me. It's old, creased and rubbed blank in places. Held together with brittle yellow Sellotape and missing its front cover, it's falling apart at the seams, but the swirling heathland contours and pale green blocks of woodland are decorated with scribbles. Cryptic handwritten codes scrawled as a teenager; circles marking the location of Roman potteries, bird nests, badger latrines or interesting trees. Faded pencil marks record rare sightings or childhood events. A montage of useful and useless memories and information.

I've had this map for a very long time and even though it's too delicate to use in the field, I still travel with it jammed above the sun visor in my van. I must have spent hours studying it over the years, but sitting back to look at it now I'm acutely aware of how localised my scribbles are. Whole tracts of land are devoid of notes and while the west sheet looks like a spider's crawled over it, the south-east of the forest might as well be terra incognita.

When I was young, a trip into the forest meant visiting a few particular parts. I had my favourite haunts and returned to them over and over again. The better I got to know these places, the more I wanted to return. I felt at home there and came to know them intimately. The joy I experienced from stalking deer, climbing trees, sleeping rough or exploring never stopped growing. My patch, such as it was, is very much the area I have spent these last few months filming. It's where I instinctively gravitate to. Knowing which view is the best at what time of day, or where the deer are most likely to come to drink, or where the tawnies have nested almost every year since I was fifteen is all extremely useful when it comes to making a film.

My favourite part of the forest was and still is the north-west quarter. This was the easiest region to reach from my teenage home in Ringwood. Despite a pressing need to keep things within cycling distance, this north-west area has always had a wildness about it that has been lost in much of the central and southern districts. Maybe this is simply because I don't know those areas well enough. Since I don't live in the forest any more and will probably never be able to afford to return, I am stuck in a kind of time warp, with this region representing a microcosm of everything I ever loved about the place as a whole. In my opinion, it is one of the most beautiful, fascinating and spiritually rejuvenating places on earth. I'm just lucky that despite now living elsewhere, I'm still able to return whenever I feel the need. Yet, even in the west of the forest, in the midst of a place I know so well that I can picture its trails as clearly as the veins on the back of my hand, there are still plenty of corners I know nothing about. So, when it was decided to include the hidden life of a mire in our film, I recognised it as an opportunity to learn more about a habitat I'd seen only as an obstacle in the landscape to be avoided. One of the joys of wildlife film-making is the opportunity to delve a little deeper. To fill in the blanks. So, turning my attention to the patches of white lying between the familiar blocks of green, I take a closer look at our options.

Unsurprisingly, there are many different types of wetland in Britain. As with the apocryphal fifty words for snow, the lines of differentiation are considerably blurred. Marsh, bog, swamp, mire, fen, carr: the specific terms often relate to acidity, water flow, water source and topography. The kind for which the New Forest is most famous are known as lowland acidic valley mires. A peaty, indigestible mouthful often abbreviated to 'valley bog', or simply 'bottom', due to their low-lying position. Bog, bottom, moor and mire are inter-changeable terms in the forest, and you're likely to see all four used within the same few inches on the map. Cranes Moor, Duck Hole Bog and the wonderfully named Slap Bottom: there are hundreds. Essentially, they're all valley mires – 75 per cent of what's left in the whole of Western Europe – and given the globally threatened status of such places, it seems only right to pay them some attention, and not only as a backdrop to curlews.

The mires tend to be largest in the south of the forest where the ground shelves more gently, but many still lurk between the steeper terraces of the north. They contain permanently waterlogged accumulations of peat, and these flooded conditions create anoxic environments where decomposition of organic matter is incomplete. Mires develop where carbon input exceeds carbon output. They become carbon sinks and despite covering only 3 per cent of the global land surface, they are estimated to hold around 500 to 700 billion tonnes of the stuff. This means that when mires are deliberately drained for agriculture or development, the pendulum swings the other way and carbon sinks become sources due to renewed decomposition. Removing the peat to burn as fuel obviously doesn't help either. In landscapes largely devoid of trees, such as parts of central Ireland and Denmark, people have had little choice historically. Even parts of the New Forest such as Cranes Moor in the west carry the scars of small-scale peat cutting and while this process of drainage and removal is obviously very destructive, it can occasionally help shed a light on remote periods of history.

The mid to lower levels of a peat bog form a kind of scrapbook of trapped minerals, pollen and animal bones. Even immaculately preserved prehistoric human remains. No bog bodies or extinct fauna have been retrieved from New Forest mires because they are protected and intact, but I'd be willing to bet they harbour many intriguing and possibly quite dark secrets. One of the most amazing things about the New Forest mires is that some of them form a direct gradient of climate information all the way back to the bleak post-glacial landscape of 11,000 years ago. A transitional time when the ice sheets were finally retreating and the tundra in this remote corner of Europe was being colonised by birch and pine. Analysis of trapped pollen, alongside the use of carbon-14 dating, reveals landscape progression from open tundra through boreal forest to broadleaf woodland.

These same mires have been here since the dawn of the Holocene, anchoring the landscape while woodland and heath ebbed and flowed around them in their perpetual tug of war to dominate higher ground. It's somehow ironic that the oldest of the forest's ecosystems is also one of its most visually unassuming. Adding a few more scribbles to my map, I'm making an attempt to choose between these blank areas.

For old times' sake, I choose a place called Soldiers Bog. Or at least that's what I think it's called. It's not actually named on the map and the only indication it even exists is a collection of thin wavy blue lines. But I know it's there because when I was thirteen, I fell in up to my waist while trying to cross it. Aptly enough, given the state of my clothes once I emerged, it skirts a place called Stinking Edge Wood and although I remember it as being a beautiful, secluded little valley, the last thing I want to do is make it deliberately harder than it needs to be or to dump £100,000 of camera kit into a bog accidentally while struggling to cross it. Looking more closely at the map, I'm relieved to notice a small footbridge located a little way upstream. Apparently, it's always been there. Embarrassing.

The sun is high by the time we arrive. Approaching on foot from the ridge above, we follow the old track down through the heather into the small valley. Stinking Edge Wood stands ragged and ancient on the opposite side of the mire and the warm air is thick with the heavy, resinous scent of bog myrtle. The sun shines silver from the languid pool to our left and a metallic glint draws my eye to an emperor dragonfly flitting back and forth over burnished water. The air is alive with insects. A pair of siskins flies up from the water's edge and a snipe breaks cover in a flurry. Pausing on the narrow bridge to watch the bird jink away up the valley, I put down the camera and lean against the wooden handrail to get a feel for the place.

Ancient places such as this have a way of making you reflect on the passing of time. It's been years since I was last here, but memories come flooding back, snippets of imagery flashing and chasing one another through my mind. Some sharp and intimate, others blurred and remote. All of them evocative. A thirteen-year-old boy perched in an oak to watch two fighting stallions: rolling eyes, biting teeth, ragged manes. Two teenagers cooking up rabbit stew on a winter's day: damp bracken, cheap wine. The pale ghost of a hen harrier floating over the heads of a father and son immersed in their first adult conversation. This last memory brings me up short. It's been almost a year since Dad and I last saw each other. Work, lockdown and skewed priorities conspiring against us. I make a vow to be more proactive once the world makes it possible again.

Turning my attention back to the here and now, I look upstream. Soldiers Bog is just the final stage of a much larger mire originating in Backley Bottom further up the valley. Seeping slowly down through alder carr, willow and mops of tussock sedge, rainwater arrives from higher ground to collect in a large spire bed of reeds before brimming over to flow beneath the bridge I'm on. Pooling again to my left, it percolates through a sieve of lush vegetation on its way downstream towards the headwaters of Blackensford Brook beyond. It's a beautiful spot fringed by a thousand shades of emerald and studded with the tiny yellow flowers of creeping buttercup. A jumble of grasses and floating beds of flowers. Beneath this seemingly haphazard, effortless beauty lie complicated patterns of water and nutrient movement.

The most fertile part of a mire is its centre. This is where nutrients from higher ground accumulate, and where damp-loving trees and shrubs like to grow. These boggy islands become further enriched by the nitrogen-fixing behaviour of bog myrtle and alder. The most nutrient-poor parts of the mire are its outer edges, home to extensive beds of stoachy sphagnum moss and carnivorous sundews.

Sundews eke out a living from digesting small insects trapped, as Darwin observed, by the 'small drops of liquor-like dew' that sparkle enticingly on their flat, fleshy leaves.

I can't pretend to know all of the plants that make up an acidic mire's community, but suffice to say it's an extensive and impressive mix. Up to 150 species have been recorded in the best New Forest sites. The list reads a bit like a magic potion, full of old English folk names and obscure references to traditional use. Plants such as gipsywort, buckthorn, bogbean, loosestrife, flote grass and spearwort. Indeed, many of the plants that call mires home were once used by people for a whole host of things. Gipsywort gets its name from its use as a clothes dye by Romanies and as a skin dye used by criminals apparently wishing to pose as Roma. Bog myrtle, known locally as gold withey, was used as an effective insect repellent, but its sedative properties were also 'said to be extensively used' throughout the forest for the purposes of 'drugging beer' in the days before the use of hops became the norm, according to John Wise, writing in 1860. The miracle moss sphagnum, referred to locally as gold-heath, was traditionally used for anti-bacterial wound dressings and 'fine brooms and brushes'. St John's wort has a well-known history in treating depression. Bogbean helps overcome insomnia; loosestrife is an expectorant. The list goes on and on and a well-established valley mire is a veritable botanical treasure trove. And we haven't even begun to talk about its fauna. Like all complex ecosystems, this mire is more than the sum of its parts and has far-reaching influence on other habitats around it.

Valley mires often form the headwaters of streams and rivers that flow down to the Hampshire coast. The Backley Bottom and Soldiers Bog system feeds Blackensford Brook, which is part of a wider catchment of tributaries and drainage ditches that wind their way east through the forest to join the Lymington river north of Brockenhurst. From here rainwater that originally fell on the heathland above Backley Bottom finally reaches the Solent, although how long this takes is anyone's guess. What's important, however, is that slow is good. Flash flooding, erosion and nutrient loss caused by fast flows of unimpeded water can be mitigated by allowing watercourses to meander and lay down deposits naturally, which is why so much effort and money is currently being invested in restoring watercourses to their natural state – a notion that will continue to grow in importance due to the calamitous effects of global warming. It's a sad irony that so much money is now being thrown at reversing the overzealous, destructive and expensive drainage projects of yesteryear.

Despite its untouched atmosphere, Soldiers Bog was renovated a few years ago during mire restoration work funded by the Higher Level Stewardship scheme. Heather bales were submerged to help raise the level of the bog and re-establish the all-important meanders. Not that you'd know if you hadn't been told, but it is clearly benefiting the local wildlife hugely. The boom-or-bust flow has been stemmed and the benefits of this habitat consistency are clear to see in the complexity of vegetation and insect life.

The dashing emperor catches my attention. Leaning forward on the handrail, I try to decide how best to capture its elemental beauty on camera. Although large for a dragonfly, it's still only 8 cm long and flies like a witch on amphetamines. A top speed of 30 mph equates to a forward momentum of 175 body lengths per second. Size for size, around 500 mph in human terms. Dragonflies don't inhabit two-dimensional planes and their ability to swerve, jink, hover, stop dead or even fly backwards is of course legendary. All of which poses a bit of an issue for anyone trying to fill frame with them through a

1,500 mm telephoto lens. Taking up a position beneath the shade of an oak tree growing by the pool, I try to get my eye in with a few practice runs.

Emperors are fiercely territorial, so the bright-blue male zipping around in front of me is on his own and seems to be following a reasonably predictable flight route from one end of the pool to the other. Running the camera at 240 frames per second rather than the standard 25 slows movement down tenfold, but obviously does nothing to slow the action in real time and I spend the next few hours panning the camera quickly from left to right, trying to keep focused on the indefatigable insect. Keeping the pressure on and not letting up usually pays off sooner or later, and by lunchtime Matt and I have a handful of shots I'm happy with.

Slowing the dragonfly down on screen reveals an endless stream of subtle movements and micro-readjustments that normally remain completely hidden. Rainbow colours arc across the huge compound eyes as it turns into the warm breeze being funnelled down the valley. Its head remains level like a gimbal, while its powerful green thorax rotates around its axis, banking like a helicopter before levelling out again. Its hair-thin legs are folded and tucked up beneath it and the long electric-blue abdomen has a slight downward curve. I can clearly see its body pulsing with the effort of sucking in oxygen through its spiracles and once in a while the wings stop beating to let the insect glide and recover.

At full speed, the wings beat thirty times a second, which isn't actually that fast for an insect, but slowed down I can clearly see each ripple and shimmer. They don't only beat up and down but rotate and flex also. There's clearly an awful lot of subtle micro-variables being continuously adjusted. At this level of detail, it's almost possible to feel the buoyant flow of fluid air and the crush of gravity on the turns. A dragonfly will experience three times the G created by the Space Shuttle during take-off, and the emperor is so perfectly designed, it has remained unchanged for 280 million years, compared to Homo sapiens' 200,000. A mind-melting length of time. The phrase 'if it isn't broken, don't fix it' springs to mind and looking at this little miracle of evolution, it's hard not to think that my own species has barely started out. In the face of such timeless perfection, humans clearly have an awful long way to go and an awful lot to learn while getting there. Dragonflies have lived in perfect harmony with the planet for all that time, while the way we treat it makes me sometimes wonder whether we are as sentient as we like to believe.

My morning's audience with the emperor has left me wanting more. Searching downstream, I discover other dragonfly territories stretching in a glittering chain down to Blackensford Brook. To be accurate, some of these belong to damselflies too and over the next hour or so I pick up a few shots of the nationally rare and excruciatingly fragile small red damselfly and, my favourite, the beautiful demoiselle: a long-legged damselfly that perches with abdomen erect and dark rounded wings closed. It is this last, fairly ubiquitous, damselfly that I associate most closely with the forest streams – a bewitching creature encased within metallic green armour that flashes and flares in the sunlight. Having staked a claim to a short section of stream, it seems content to remain perched on sprigs of water mint, turning slowly to face the breeze like a jewel-encrusted weathervane. Its heavy dark wings beat more slowly than those of a dragonfly and they whirl and twist in complex, hypnotic patterns.

It has turned into a very hot day and the fresh scent of myrtle has given way to the waft of dry earth and pony dung. Retreating back to the shade of an oak for lunch, I catch sight of a large raptor wheeling over the ridge above us. Its long paddle tail, huge yellow feet, powerful chest and heavily barred wings are unmistakable: goshawk. A large female riding the bumpy heathland thermals. Having gained height, she peels off and powers her way up the valley towards Bratley Wood. I can count on the fingers of one hand the number of times I've seen a gos out in the open like this. She's clearly on a mission and has probably been driven out to hunt in the sun by a nest full of rapacious chicks. The sight of her reminds me that it's still business as usual out there in the woods and that our own goshawk chicks will be leaving the nest for good any day now. I was there only on Monday, but with their new flight feathers growing at the incredible rate of 1 cm every day, now is not the time to get distracted. Come what may, I need to be back in that hide the day after tomorrow.

Shadows reach out from the wood as the afternoon wears on and one by one the damsels and dragons stop flying. A water-cooled breeze steals the heat from the day, and I hear the dry-earth thud of ponies approaching through the trees behind us. Emerging from the eaves of the wood, a large multicoloured herd of thirty fan out and saunter towards the wooden bridge, coming down to drink. A young bay steps into the pool up to her belly. Lifting a front leg, she paws the water playfully, splashing herself and others while tossing her head and whinnying with obvious delight.

It's been a hot day and I can feel the sun in my muscles as we carry the camera kit back across the heath. Re-entering the realms of signal, I hear Matt's phone ping and he shows me a photograph of Bournemouth seafront on social media. Taken around lunchtime, it shows a crowded beach chock-a-block with holidaymakers with barely enough space to lay out towels between them. An estimated 500,000 people descended on Dorset today, some from as far away as the Midlands. Great if it wasn't for the fact that we're in the middle of a pandemic. We still don't understand the true effects of such outdoor contact and the police have labelled it a 'major incident'. It's hard not to see it as yet another depressing sign of our nation's diminishing grasp of reality in the face of a Covid death toll now above 43,000.

Thursday 25 June

Our second day down at the mire. It's set to be another hot one, which means the dragonflies will be active and the bog simmering with life. Having set up again beneath the oak tree on the edge of the main pool, I pour myself a cup of tea and take a stroll downstream. There's something fascinating going on everywhere I look, which is how Matt and I soon find ourselves lying on our stomachs scrutinising the mating behaviour of semaphore flies. Clouds of these tiny metallic insects are clustered around the mire edges, engrossed in one of the most complex courtship rituals known. Not that either of us knows what they're up to at first, but having sauntered off on the pretext of going for a pee and finding a patch of reception with which to steal knowledge from the Google gods, I return to Matt and deliver an erudite bogside lecture on the complex mating rituals of *Poecilobothrus nobilitatus*. Not that he's fooled. But he's polite enough to listen and soon we're both fully engrossed in the drama playing out on the surface of the water barely inches away.

Viewing the world of insects through a macro lens is one of life's delights. To be transported into such an alien landscape is like stepping into a parallel dimension. Each visit leaves you craving more and is so utterly engrossing that it's often a real effort to pull your-self back out. At this tiny scale everything we take for granted in our own reality seems turned on its head. Water is solid and things move a hundred times faster. Smooth surfaces are fissured and even light seems more volatile. I must appear so huge, slow and clumsy through their bulbous compound eyes that the insects barely seem to notice me. Both sexes are green and yellow, but only males have dark wings tipped with white. The females calmly go about their business, munching mosquito larvae and perching on pondweed, but the males are all over the place. Chasing each other and indulging in extrovert aerobatics in an effort to persuade the females to forgo their breakfast in favour of mating.

It's a chaotic scene, but after a while patterns start to emerge. Having chosen a female, the would-be suitor lands on the water and turns to throw her a look: the 'blue-steel' moment, designed to grab her attention. If she stays put, he then catwalks casually towards her, holding out his vibrating wings. If she likes what she sees enough to ignore her squirming breakfast, he'll then hover in front of her and do a few barrel rolls. At this point he may become distracted by rival males and see them off, but he quickly returns to his wooing, and if the female has waited, they'll take to the wing together before finally mating. The whole thing happens so fast it's hard to keep tabs and is over within thirty seconds. It's a fantastic bit of courtship behaviour strangely reminiscent of the complex dances used by the birds of paradise I've filmed in Papua New Guinea. Not as flamboyant, but just as engrossing.

As the sun rises higher, I notice iridescence shining on the black water snared between reeds around me. It coats the surface like an oil spill and refracts sunlight in shards of silver and blue. Pond skaters drift slowly on hidden eddies and the surface sheen is cracked and split like a satellite image of arctic ice floes. Like a fractal, the patterns get more intricate and beautiful the closer I look. Matt asks whether it's caused by pollution, but this is one bog-related phenomenon I know well from living near the Somerset levels. What at first glance looks like spill from a chainsaw is actually perfectly natural and caused by one of the oldest forms of life on earth.

Leptothrix bacteria live in colonies that grow in size as bog water grows stagnant during dry weather. Evaporation naturally increases the relative levels of water-saturated heavy metals, such as iron and manganese released originally from the decomposition of organic material. Leptothrix metabolises such minerals, resulting in thin layers of oxidised metal accumulating on the water's surface. These kaleidoscopic layers are actually solid, which is why they appear to shatter like stained glass when poked with a stick (unlike oil, which flows). This thin mineral crust can even be strained off to leave clean drinking water behind. Such bacteria have been around for not just millions but billions of years (think primordial soup) and are an essential part of our ecosystem. By making trace metals available for plants to absorb, they are a cornerstone of life on which entire nutrient cycles and food chains are based. Not bad for an organism so humble that most of us notice them only because their waste looks pretty. Yet another example of why the blank spaces on a map shouldn't be ignored.

The mercury is rising, and the air wobbles and collapses around us. Thermals suck cooler air across the heath, but beneath the trees it feels close and stifling. Today's the hottest day of the year so far, and while the news feed shows more intense scenes on Bournemouth beach, we've yet to see another soul, so continue filming as if the rest of the world doesn't exist. Today is our last day here. It's been a welcome place to hide out, but we need to refocus on goshawks before the chicks leave the nest for good. Before we leave though, this unassuming corner of the forest has one more treat to share.

A pair of meadow pipits has been flitting around for most of the day. Collecting beaks full of leggy insects, they've been flying back and forth between the pool and a copse of alder. Presuming they have a nest in the trees, I got a few shots of them foraging, but left it at that. As I'm watching them, half wondering what to film next, a large long-tailed bird breaks cover and flies down the valley through the shimmering air towards us. It's heavily barred grey-and-white plumage dupes me into thinking it's a sparrowhawk, but as the approaching mirage sharpens, I realise with a start that it's a juvenile cuckoo. The penny suddenly drops, and I realise what the meadow pipits have been up to all this time. They haven't been feeding chicks in a nest at all and grabbing the camera I try to predict where the young cuckoo is going to land. An old holly seems the likeliest place and I get the camera lined up just as it makes contact.

Still looking strangely hawk-like in its banded plumage, it leans forward, huge yellow beak gaping wide. The fledgling starts to beg, turning its yellow mouth in search of its surrogate parents. Its high-pitched repetitive calls drift across the wet ground between us as its right wing opens, trembling with anticipation. A few seconds later a tiny pipit lands on its shoulder before reaching forward to place its entire head inside the chick's cavernous mouth, like a lion tamer at the circus. The cuckoo gulps forward as if to swallow the tiny bird whole, but then jerks back and the pipit flies off, its beak now empty. Within ten seconds, the other parent arrives and repeats the process. And on it goes, over and over again: gawk, squawk, gulp; gawk, squawk, gulp. And while it's truly amazing behaviour, I can't help feeling sorry for the birds who are clearly being driven to exhaustion by their chick's relentless demands. The two pipits seem trapped in a never-ending attempt to satiate the insatiable, as if being punished for some unknown transgression. A few minutes later, the cuckoo flies on to another tree, and I notice that it's always the right wing that trembles, never the left. And it's always the cuckoo's right shoulder on which the pipits land. Little mysteries that will probably never be explained, but then the cuckoo is a very mysterious bird.

I can't think of any other British bird with a more mercurial reputation or fascinating folklore. Once believed by some to turn into sparrowhawks during winter, and by others to hibernate in the stumps of trees only to fly out of the embers like a phoenix should the tree be felled and burned, its spring arrival has always been associated with wealth and fertility. It's also been associated with infidelity, sexual wantonness and female deceit. It seems we enjoy heaping baggage on these birds, but while the surreptitious parasitising of other birds' nests provides obvious material for end-less allusion and allegory, one of the things that fascinates me most is the bird's effective biomimicry. By being deceived into believing I was watching a hawk, I was being manipulated by the same tactics that evolved to deceive the small birds being parasitised. By disguis-ing itself as a songbird-specific predator such as the sparrowhawk, the cuckoo creates distraction and buys precious seconds to lay her egg in the nest of a bird trying to avoid her. Why the male cuckoo is also dressed like a raptor is a mystery to me, but maybe the more confusion the better and it may help to draw the focus of mobbing behaviour away from the female. In addition, the disguise might also reduce chances of the cuckoo being predated by actual hawks themselves.

Not all songbirds are targeted though, and British cuckoos seem to favour dunnocks, reed warblers and of course meadow pipits. Not only this, but they are known to return to the same places year after year and these locations are influenced by where they themselves were raised, as is their choice of target songbird. A cuckoo raised by pipits on Bratley Plain will return to Bratley Plain to lay its own egg in the nest of other pipits. On top of this, the egg the cuckoo lays is coloured and patterned to blend in with those of the host species traditionally targeted by her ancestors, including her mother, and are held within the female's body until the very last moment in order to synchronise its hatching with the host birds' own clutch.

All in all, the cuckoo really is an amazing piece of work. It doesn't stop there, because despite having pride of place in British folklore, it actually spends most of its life elsewhere. Revealingly, the bird we've adopted as a symbol of British springtime spends only a few months of the year with us. Most of the New Forest cuckoos arrive mid-April but are gone again by August; many adult males have even gone by early June. But where? Not hollow tree stumps, that's for sure. And why are fewer and fewer returning each year?

Like a twenty-first-century eco-cliché, the basic facts are once again sobering and reminiscent of the plight of the curlew and so many other species. England has lost 68 per cent of its cuckoos within the past twenty-five years and while slightly lower losses in Scotland and Wales pull the overall national average up to around 50 per cent, this is still appalling and bewildering. Bewildering because cuckoos that manage to get here stand a pretty good chance of breeding successfully. So, if the decline isn't due to obvious breeding pressures, as it seems to be for the curlew, it suggests something else is happening during migration. In 2011 the British Trust for Ornithology launched a radio-tracking programme to try to discover what.

Data from forty-two tagged male birds between 2011 and 2014 showed that British cuckoos either took a western route from the UK down through Spain, Morocco and West Africa, or an eastern route via Italy and Libya. Both routes involved an astonishing sixty-hour non-stop crossing of the Sahara and the ultimate destination for all forty-two was shown to be the Congo basin, an area I know and love. That's quite a journey – mostly completed at high altitude at night. Surprisingly, though, most fatalities occurred in Western Europe, during what I'd assumed to be the less challenging leg of the journey. Large-scale habitat loss in Spain is considered an issue, but also the reduction in available food in the breeding grounds of south-west Britain, the regions where those birds using the western migration routes come from and return to. Cuckoos eat insects but specialise in caterpillars too hairy to be eaten by other species. They fatten up on these prior to migration. It seems that regional reductions in British insects, particularly the Lepidoptera, are having an influence. Part of a wide-scale depletion of the British countryside, particularly here in the lowlands. We may live in a green and pleasant land, but if it becomes an empty stage devoid of insects then everything else in the food chain inevitably suffers. Protected non-arable areas large enough to encompass their own water catchments such as the New Forest should in theory experience less damage. Even so, it would be naive to assume that national parks can survive as intact ecosystems regardless of what is going on out there in the larger landscape around them.

By dusk, the landscape has mellowed into a wash of soft greens and purple. The ponies have returned for an evening drink and knowing that tomorrow I'll be back with the goshawks, I go for a short stroll on my own. A grey wagtail flits and bobs along the stream ahead of me and wood crickets call softly from the bracken. Dusk is filling the valley and a stock dove calls in the distance. Sitting back on the gnarled roots of an ancient beech, I listen to the sound of the breeze murmuring through dry leaves and the soft creaking of old timber. It gives me great comfort to know that places and moments like this still exist. Especially now, in the face of so much threat and uncertainty.

Perhaps my natural tendency has always been to run and hide. Maybe that's why I've chosen the job I have. And while it's easy to pass comment about those on whom we've heaped our hopes and fears for the future, I realise that, as one of the multitude, I'm in the enviable position of not having to stick my neck out. But I'd be lying if I said I wasn't nervous about where all of this is heading.

Britain feels divided at the moment. The world feels divided, with whole continents fragmenting and drifting away from each other as the death tolls rise. Divisions sow distrust. If left to fester, I worry about this morphing into something else entirely, and a destabilised society is vulnerable to all sorts of scary issues that would've seemed inconceivable a short time before. I've experienced this first-hand in central and west Africa. The toxic cocktail of division, distrust and fatal disease now brewing on the world stage worries me to the point where I struggle to listen to the radio, let alone watch TV. Confusion, fear, anger and suspicion are gathering pace, and I can't help thinking that things are set to get a whole lot worse if we can't get on top of it before winter comes. I guess only time will tell what we've done right and wrong during this pandemic, but we really should be getting our act together better than we are. On the other hand, maybe the people on Bournemouth beach have got it right after all. Why wait for the creeping inevitability of bad news? 'Rosy days are few', so why not grab a few hours in the sunshine and let the chips fall where they may?

A tawny owl calls from the depths of the wood behind and a snipe starts to drum, filling the still valley with vibration. Shadows are rising and the pastel sky is soft. The last rays of the setting sun linger on the tops of firs on the ridge above and if I don't rouse myself to head home now, I'll be spending the night here.

Friday 26 June

Hunched like a vulture, the male chick leans forward, scrutinising every twig, leaf and flake of bark in the wood. The occasional flick of a nictitating membrane draws my eye to his, where the dappled sky is mirrored in pale grey crystal. Shaking the reflection away, as if in irritation, he lowers his head to preen. Barely any wisps of white down remain. Bronze spearheads fleck his breast and the tips of his long tail feathers are losing their immature fringe of white. Lifting his head, his eyes lock onto something moving quickly through the trees towards the nest. His rising hackles tell me it's an approaching adult. Almost tripping over his own talons in his eagerness to turn, he shrugs open his wings and starts to call – a reedy high-pitched screech that immediately alerts his sisters perched in other trees close by.

The nest has remained empty all morning, but with a flash of lead-weighted grey, the adult goshawk lands square in the middle of it. It's the female. Huge, sculpted, perfect. A squirrel lies pinned beneath her, its soft matted fur contrasting with her powerful yellow feet. It's clear she's enjoying hunting again after those long, confined weeks. She's in fine yarak – tip top hunting condition. A heartbeat later she slides off the nest, dissolving. A beat after that, the first chick arrives to take possession of the food left behind. It's the older of the two sisters, except she's not a chick any more. In the space of two short days, it seems she's finally grown into herself and overtaken her brother. Although she staggers on landing, she's coming on quickly and it will only be a matter of days before she's capable of sustained flight. A moment later her younger sister arrives, followed by the male, who lands with a bounce, ready to take control of the squirrel. But the power balance has shifted, and his larger sister ignores him, mantling the corpse and tugging it apart with barely a glance at her siblings. Wedging his way between his two sisters, he manages to hold on to second place while waiting his turn, but it's clear that he'll soon be relegated further.

Having eaten her fill, the bird hops up to her favourite branch to preen, then gradually works her way from tree to tree until she's perching 20 feet higher, 50 feet away to the left. The male too eventually goes his own way in the opposite direction, leaving the youngest sibling to scavenge what she can from their leftovers. There doesn't appear to be much, so she sits down in the hope that more will arrive soon.

The hours slide slowly by and the day grows hot. The solitary chick lies panting in the heat and plagued by flies. I wonder why she doesn't move into the shade. By late morning, her frustrations get the better of her. Rising to her feet, she emits a flurry of begging calls, then scowls at a larch cone attached to a twig on the edge of the nest. Quick as an adder, her left foot strikes out to grab it. With wings and head held back in attack mode, she wrenches the cone from its branch, picks it up in her hooked bill in a strange parody of the robin's head. For a moment she looks as if she's going to eat it – but then spits it out in disgust and sinks back down on her haunches to sulk.

Then, just as it seems she's going to stay hungry for the rest of the day, there's another flash of grey and her father drops something onto the twigs in front of her. Looking small and delicate compared to his hulking daughter, he barely alights before turning. I catch one final glimpse of his otherworldly eyes, then he too disappears and the chick is once again alone. I can hear the other two clattering their way back to the nest, but they're too late. A small scrap of brown dangles from their sister's beak and, just before it's devoured, I recognise the tiny pink feet, long tail and large round ears of a wood mouse. Limp and pathetic, it's devoured in the blink of a pale grey eye. It's not much to stave off the hunger of a growing goshawk, but she seems content enough and is soon branching her way up the larch above her. As I enter the closed canopy, a dark flurry of wings tells me she's dropped down to a favourite roosting tree nearby. All three chicks have now melted into the background and for the first time in three months I'm looking out on an empty stage.

I stay until dusk. The nest remains deserted and as much as I'd love to return again tomorrow and the next day, to catch one last glimpse of the birds, I realise that this is it. From a filming point of view the season is over. The filming hide will remain here for a while, its empty windows overlooking the empty nest. Once removed, the only thing left to show that this was once home to a family of wild English goshawks will be the last few scraps of down fluttering gently in the breeze.

Epilogue

February 2021

IT'S 2 A.M. AND I LIE SWEATING IN THE DARKNESS OF THE bedroom. My pillow is soaked and moisture pulses in the notch at the base of my throat. Bad dream. My father died three months ago following a short, bitter struggle with cancer. Ten heart-rending weeks lay between diagnosis and death. Shock, anger and grief have all melded together into a Frankenstein's monster of emotion that lurches through my mind wreaking havoc while I sleep. It's hard to keep a lid on it, but the middle of a third lockdown is not the time to drag it all out. My feelings are still too raw and unpredictable. My wife and I talk when we can, but life goes on and the daily routine of homeschooling three boys while working from home is more than enough for us to be dealing with right now. So, slipping out of bed, I tiptoe to the front door and step out into the crisp beauty of a moon-frosted night. I suck the sharp air down into my chest and stare up at Orion.

Going back inside, I cook up some eggs, brew a pot of coffee and by 5 a.m. I'm locking the front door and scraping the brittle ice from my windscreen. Ten minutes later and I'm following the old familiar road back down to the forest.

I'm smiling up at a woodlark singing his heart out. His pale breast glows pink in the rising sun and his translucent wings tremble against the brightening sky as the melody pours out like water. A fluting, heart-lifting song so pure, soft and sweet, it embodies the very essence of early spring. The perfect tonic to a troubled night. He parachutes back down to earth in a graceful arc, short tail cocked. A moment later a Dartford warbler breaks cover, skimming the scrub to perch in the gorse off to my left. He's handsome, but his mumbled song can't compare.

I'm standing on the shoulder of an open ridge that rises up from a sea of trees at the centre of the forest. In front of me, the ground drops away into a shallow, shadowed valley crowded with birch. A solitary hawfinch perches in the gossamer of their topmost branches and a lesser spotted woodpecker drums from somewhere in the distance. Beyond this, the forest rises up into the morning light to cover ridge after ridge in a vast canopy of winter-washed silver. The domed canopies of oaks heave up against jagged islands of fir, while far away on the horizon a line of redwoods towers above everything else. This is the forest's heartland, and one of the largest tracts of mature woodland anywhere in Britain. Between me and those distant sequoias lie fifteen square kilometres of mixed broadleaf and conifer. Beyond the horizon lie many more. It's a rare view and one that fills me with wonder every time I come here.

It's also one of the best places in the UK to watch goshawks displaying. My vantage point looks out over six breeding territories, including the very first established in the forest back in 2001. Tea in hand, I know that somewhere out there right now a dozen or so goshawks are starting their day. Not that you'd know. It's not in their nature to reveal themselves casually and for 300 days of the year, you could come here and remain none the wiser. But for a precious few weeks at the end of each winter, the phantom throws off its cloak of secrecy and takes to the sky.

It's the males that hold territories during winter. The home area contracts, but he will maintain a presence, killing only what he needs in order to make it through to the next breeding season. A wood pigeon every other day or so, or something of the like. But as winter loosens its grip and the females drift back in from their wider-ranging winter grounds, males take flight to welcome their partners home. Goshawks aren't monogamous in the true sense, but most females instinctively return to the previous year's territory. If the same male happens to be resident, and if she was happy with his performance last time around, then they hook up to do it all over again. The process of re-establishing bonds after winter also helps ward off potential rivals, including first-year juveniles looking for a territory of their own.

The first blue-sky days of February are best for displays. Both sexes can be seen soaring and wheeling together and males hurl themselves across the sky in great roller-coastering climbs and nose-dives. Having committed to each other, they then patrol together, using deep, exaggerated wingbeats and fanned-out tail coverts to emphasise their claim. These conspicuous puffs of white stand out for miles, helping set goshawks apart from buzzards.

Goshawk displays are inherently ephemeral but pick the right day and you'll be rewarded. Pick the wrong day and you're left second-guessing why the birds aren't showing. It can be a frustrating experience, but there's usually something else to watch in their absence. So, having set up my camera beneath a solitary birch on the heath, I raise my binoculars to begin the vigil.

The day wears on, but apart from a few buzzards and an empty Thermos there's precious little to show for it. A steady stream of birders and photographers passes by, mostly on the hunt for a local grey shrike, but for the most part I'm left to ponder the sky in silence. Then, mid-morning I hear the sound of a 4×4 approaching along the ridge behind me. I turn around to see a familiar green Hilux emerging from behind a thicket of gorse. It rolls slowly down the track and pulls up 20 metres away. Andy gets out, holding his packed lunch. It's only mid-morning but it seems he was up early too, and the sight of his food has me reaching into my rucksack for mine. He's come to help me spot the goshawks. A second pair of eyes goes a long way and I'm grateful.

'I'm not sure I can cope with any more of this bloody deskwork,' moans Andy as he sits down in the heather next to me. I can't say I blame him, not on a day like this.

He's looking paler and more tired than when I last saw him. We both are. The long, hot, energised days of last summer feel like a world away and two more lockdowns either side of a wet, dreary winter haven't helped. Nevertheless, clear blue skies go a long way towards lifting the soul and, as if on cue, my first brimstone of the year flits by.

I ask Andy how long he's got left until he retires – until that desk can be left behind.

'Not sure. I'll either go this September, or next, depending . . .'

He takes a bite of sandwich, leaving me to wonder who will take his place as future head of wildlife management in the forest. There's part of me that would jump at the chance myself.

'Will your successor be chosen from within Forestry England's ranks?' I ask.

'Probably. There's just so much to learn though and you have to be prepared to make difficult decisions. I haven't made many friends among the forestry lot over the years, that's for sure.'

I presume he's referring to the restructuring of tree-felling programmes in order to protect rare breeding birds, including goshawks, who often set up home in the biggest, most commercially important stands of fir in the whole forest.

I point out that he wouldn't be doing his job properly if he hadn't ruffled a few feathers along the way. And I mean it. It's a dilemma that lies at the heart of wildlife management. At what point should ecosystem restoration give way to commercial or cultural interest – or vice versa? It's an issue that's particularly important in an area so steeped in human history as here. Ultimately, just how compatible are concepts such as commoners' rights, unfettered public access and commercial logging, with the encouragement and protection of biodiversity?

For me, the successful return of species such as goshawk, pine marten and otter raises the question of whether other lost species should be actively reintroduced as part of a more conscious wilding programme. The concept of rewilding is very popular at present, and I'm all for it. So far, aside from the reintroduction of sand lizards in the 1990s and white-tailed eagles on the nearby Isle of Wight in 2019, there haven't been any other officially sanctioned programmes in this area that I know of. I'd love to see beaver, elk and boar back in the forest, for example. Hell, I'd love to see lynx back here too but realise I'm not exactly a fair representative of opinion on that one. Are such aspirations of rewilding even viable for a landscape so beset by human interest, agenda and influence?

There's no avoiding the fact that this is not (and probably hasn't ever been) a wild landscape, and any temptation to consider it as such may simply reflect our own shifting baselines of what 'wild' really means, or a fundamental lack of understanding of how such landscapes have been managed in the past. But on the other hand (asking at the risk of being shot for sedition), do the woods even need managing anyway? Our ships are now made of steel, there's no requirement for pit props in coal mines any more and most of our construction timber is imported, so surely there's a case for simply letting these woods take care of themselves. I mention this to Andy and immediately realise I should have thought it through a little more first. He explains that positive management for timber production is why we've got these woods here today in the first place and to think that you'd get this by just abandoning things is a bit naive.

'It's not as plain sailing as people think, just to let nature do its own thing. These woods have been managed for centuries and centuries and even though we talk about them now as being remnants of the wild wood, that's simply not true: they've always been managed. With the deer, ponies and cattle that have been on this forest all that time, you cannot grow trees here without some work or protection going into that. It's a balance, see?'

Having already interfered in the first place by introducing domestic livestock, felling trees, damning watercourses and all the rest of it, he tells me that 'we can't then just walk away'.

It's this last comment that strikes to the very heart of the unique issues faced by the New Forest and, since we're talking, I ask what he thinks the lasting benefits of that first lockdown might have been. Andy shifts.

'Well, to be honest with you, it was so short-lived. It had the potential to be really good, but then it just exploded in our faces and the damage that was caused when we came out negated any good that was done during it.'

I flinch at the memory. 'It was pandemonium, wasn't it?'

'It was,' he replies. 'The forest was full of people who have no idea how to treat the countryside and left their litter, their rubbish; they just parked where they wanted, walked where they wanted and had no respect for anything. It was an absolute nightmare.'

Nevertheless, Andy tells me that holidaymakers and visitors to the forest are usually very open to learning something new about the area, that it's part of the experience for them. He says that in many ways he finds them more agreeable than some of those who live here. So this is in no way about keeping people out – God knows we need more people connecting with nature, caring about the natural world and fighting its corner – but, in Andy's eyes, this was something new.

'People couldn't go abroad; they couldn't go to holiday camp; they couldn't go to any of the attractions. They couldn't go anywhere they would normally go. But they *could* go into the countryside, so that's what they did.'

I feel myself growing sanctimonious at the unpleasant memory of those chaotic days. The levels of self-entitlement displayed by some people beggared belief. I have to remind myself that I had the enormous luxury of working outdoors throughout the whole lockdown period, of freedom in the forest. Was it any wonder that people needed to let off steam?

I think back to what the forest was like thirty years ago, when I still lived here. It definitely seemed a lot quieter back then and I guess modern realities have now caught me by surprise. I'd allowed myself to be duped into believing that the peace and tranquillity the forest enjoyed during lockdown could last. Andy explains that one of the things he and his team have been discussing is the ongoing recreational strategy for the future. What happens if the numbers we've seen this year become the norm?

'Unless that's controlled in some way shape or form, we will not have the wildlife here. Without a doubt. I think it will still look a special place, but I don't think it will hold as much in the future as it should do because public pressure will have pushed it out.'

As he's talking, Andy raises his binoculars to check something. 'Gos!'

Snapping out of it, I get on the camera and lock on to a handsome silver-clad male powering himself across the canopy below us. I see his deeply muscled chest, long square tail and yellow feet clenched up out of the way. He's keeping low and it's obvious that this isn't a display flight. He's hunting. And from the way other birds scatter like wind-blown leaves before him, they know it too. Then, just before he slides below the surface of the canopy, he pulls back and beats his way higher in a smooth arc that takes him away from us. A moment later, he dips again and vanishes back into the trees, leaving nothing more than a few pigeons still going hell for leather towards the horizon.

The excitement over for the time being, Andy puts the lid back on his lunch box and gets up out of the heather.

'Right, I'm off to do some real work.'

But as he's closing his car door, I suddenly remember another thing I'd been meaning to ask.

'What about the curlews? How did they get on last year?'

'We did a comprehensive monitoring of them, so I can tell you exactly how they did . . .'

'And . . .?'

'Forty-six pairs attempted nesting and three chicks flew the forest.' He closes his door and starts the engine.

I'm thunderstruck. I was hoping for some good news. Hoping for some tangible sign that lockdown had benefited the forest's wildlife after all. Curlews routinely lay four eggs, which in fact means that less than 2 per cent of these embryos made it through to fledging. A quieter forest undoubtedly helped at the start, but that was before it was swamped. It's bad for the curlew, for sure, but it is also really bad for us. When will we learn that conservation begins at home, not in the world's jungles, and that it is not always convenient or free from personal sacrifice? Even if such sacrifices are as small as keeping a dog on a lead when asked. Andy's news makes me rethink once more any misplaced, romantic notions of how lockdown might have benefited our nation's wildlife. I'm sorry, but goats wandering through the deserted streets of a Welsh town just doesn't cut it for me.

In some ways, the post-lockdown debacle was perhaps inevitable: a situation compounded by centuries of legislation designed to disenfranchise and separate the British public from the countryside. Between 1604 and 1914, more than 5,000 acts of enclosure were passed by Parliament, effectively stripping away public access to common land (and the right to make a living from it), the end result being that 80 per cent of us now live in cities and towns, with free access to less than 10 per cent of our homeland. A situation set to deteriorate further in the light of the 2026 deadline for registering unmarked public rights of ways, incidentally.

With such restricted options available, it's perhaps no surprise that so many people ended up corralled into the forest, vying for parking spaces and ice cream. For many, our national parks are the only viable easy-access option left, and a lot of the friction witnessed here and in other places such as Snowdonia was the result of too many people in search of something beyond the parks' capacity to provide. Ultimately, with half of the land in England owned by 1 per cent of its population, how can we be expected to know how to behave in the countryside when we aren't able to learn the dos and don'ts from a young age via easy access? A vicious cycle of alienation and negative fall-out that becomes a self-fulfilling prophecy. It's a simple equation: people can't be expected to care if they don't feel they belong, and I would argue that our nation's strangely medieval attitude towards land ownership and access doesn't help the situation.

I spend the rest of the afternoon watching the sky in the hope of filming another goshawk, but it's strangely quiet up there. The local birds should be displaying wing tip to wing tip, but apart from a few distant buzzards, the heavens remain eerily empty. As the sun sets in a dusky orange haze, I turn around to watch a huge moon rise over the gorse behind me. Its open, silver face brings back memories of vast flocks of curlew taking flight over a moonlit Bristol Channel – but that was twenty years ago, and it seems that the fortunes of goshawks and curlews have waxed and waned in equal measure since then.

The next morning, I find myself gravitating towards the north of the forest. I'm heading to a high point from where I can look south over the wood containing the nest I filmed. I'm hoping the adults are there again and, in the absence of any luck yesterday, I figure it has to be worth a go.

Half an hour later I'm driving through a block of trees lying to the north of goshawk wood when I notice a handwritten sign screwed to a forestry gate: 'Covid Fear = Mind Control'. Maybe I'm missing the point, but it seems a strange place to encounter such delusion, and – as I remove it from the gate – I can't help wondering what the families of the 122,000 people now known to have died in the UK alone would think about it. Having driven through, I leave the van on the track, and scale a heather-clad ridge to take the lie of the land.

I'm on the northern slope of a valley that sweeps down to a line of willow growing on the banks of a meander. Beyond this, the ground rises up and disappears beneath the impenetrable eaves of goshawk wood, the place I know so well. Having not seen the wood from this angle before, it takes me a while to get my bearings, but I eventually pinpoint the location of the hidden nest and recognise the familiar tall firs standing on the ridge directly above it. Halfway up the slope, indistinguishable from thousands of others, stands the Douglas fir that held my platform.

I set up the camera and start scanning the skyline. It's going to be another beautiful day. And, as there was yesterday, there's a woodlark singing right above me. Unlike yesterday, there's also a skylark floating even higher in the blue sky beyond. The two melodies wrap around one another, drifting down through the still air. The day grows warm, and a male hen harrier arrives, floating low over the heath on long, languid wings. I watch his round owl-like face turning from side to side as he listens for the rustle and squeak of prey. His arrival takes me back to a misty December day thirty years earlier when I saw my very first harrier on this exact same spot. Different bird I'm sure, but there's something reassuring about his presence, even if the curlews that once nested here have long since disappeared.

Noon passes, and a flock of pigeons rises from the wood like a puff of grey smoke. Where there's smoke there's fire and sure enough, I see the sleek shadow of a goshawk skimming the trees towards them. It's a large female and the pigeons have seen her. They scatter in all directions, fleeing for their lives, but she locks on to one and plunges down through the canopy after it. Not many raptors are capable of following through on a stoop like this. For most, the tangled web of a forest canopy is a dangerous, impenetrable barrier. Even a peregrine wouldn't try to enter it at speed. Yet – as we've come to see – goshawks aren't like other birds, and once again I'm reminded of their supreme adaptability and almost reckless determination when hunting.

I imagine her ricocheting down through the branches. Her body twisting and rolling; her broad wings flaring and folding as she accelerates, brakes and squeezes through the gaps. Her long tail flicking open and shut as she occupies the liminal, dangerous zone between lift and stall. I imagine her straining to close the gap between herself and her target; long, powerful legs stretched out as her talons rake the air. I sense the pigeon's terror as it feels the hawk's grasping weight and can almost see the air spilling from crumpled wings as the woodland floor rushes up into blackness.

An hour later she reveals herself again, this time perched on the exposed branch of a tall fir on the ridge. I have no idea whether she made her kill or not, although her bulging crop indicates success. Her relaxed, prominent position also suggests she's the resident female. *Our* female. And a shiver of excitement runs through me. I study her closely. She's in good condition, sitting upright with her long tail hanging below the branch. She turns towards me and her silver breast flashes in the light. Her crest is raised and, although she remains still, I can tell that she's paying close attention to the bird traffic over the wood below her. Her manner is relaxed, confident and commanding, the very epitome of Ted Hughes's 'Hawk Roosting' in the top of the wood, holding creation in her foot. And it's with a start that I realise this is the first time I've seen her out in the open, away from the nest. It's obvious that she's just as at home in sunlight as she is in shadow. She is such a mercurial bird, unknowable and unknown. As wonderful and intimate as my time with them at the nest was, it was still only a vignette, the briefest of windows on their world. In the past two hours, I've learned more about her behaviour beyond the nest than I did during 400 hours of filming and it makes me smile to think of her perched in the sunlit breeze above while I crouched inside my dark, stuffy hide for hours on end, eagerly awaiting her return.

It's also a shock that it won't be long before she's back on eggs again. It feels like yesterday that I was filming her sulk in the rain while incubating her last clutch. That was already eight months ago, and the thought brings me up short. Almost a year if I think back to my first visit to the nest last March. A lot has happened since then. My throat clenches and my eyes well up as I think of my father.

The shadows are lengthening and it's obvious I'm not going to see any displays today, so I pack up and load the van. Still, there's one final thing I need to do before I leave the forest.

My van rolls to a halt beneath the low branches of familiar conifers and I step back into the watchful silence of the wood. Nothing's changed. Out there, the sun is shining; in here it's still winter and the chilled air is laced with the scent of resin. Out there is England; in here it's somewhere much further north and I feel the hairs on my neck rise in anticipation. The sensation makes me smile and I step quietly over the ditch of broken branches to walk slowly uphill towards the nest tree. The tall, straight columns of fir slide apart to reveal glimpses and by the time I arrive at the larch I'm reassured to see the nest is still there. It's survived the winter gales and, what's more, a fresh layer of branches has been added. By the end of June, it had looked ready to disintegrate, but the parapet has since been rebuilt and a hint of green tells of fresh sprigs woven into the lattice on top. It's still too early to find moulted feathers and I don't notice any splashes of dung, but the signs are obvious. The tantalising promise of new beginnings hangs in the air.

Leaving the goshawks to it, I retrace my steps back downhill, but just as I reach my van I hear a familiar sound in the distance. The warning rattle of a wren. The cutty has seen something, and I bet I know what it is.

ACKNOWLEDGEMENTS

This book wouldn't exist if I hadn't had access to the forest during such an extraordinary year. I owe a debt of gratitude to the team at Big Wave Productions. Their determination to keep going was super-human and my particular gratitude goes to Sarah Cunliffe (CEO/Director) for inviting me on board in the first place, Vanessa Tuson (Head of Production) for keeping the show on the road, and Emma Ross (Director) for her support and positivity. Thanks also to Rhiannon Burton who has done a fine job of editing.

The New Forest is world-class but remains surprisingly unknown outside of the UK. The fact that the executive team at Smithsonian and Terra Mater had faith in its appeal to an international audience speaks volumes. I would like to thank Tria Thalman, Sabine Holzer and Martin Mészáros for their commitment and belief in the team's ability to get the job done during such challenging times.

My sincere thanks to Sarah Rigby, Publishing Director at Elliott and Thompson for seeing the potential in my sample chapters and for taking the time to chat it all through prior to offering to publish. Sarah's interest in the story and her obvious love for the British countryside spoke volumes and reassured me that E&T was the best home for this book. Huge thanks to Pippa Crane and Jill Burrows also, who alongside Sarah helped sort wheat from chaff and turn my field-observations into readable text, and to Alison Menzies for her continued help and support with publicity. It's been a pleasure working with you all.

Thanks also to Trevor Dolby, friend and literary agent at Aevitas Creative Management, for being such a generous source of wisdom and encouragement. This book owes a lot to you, not least its title!

*

I am also hugely indebted to the Forestry Commission, Forestry England and the New Forest National Park Authority. In particular, Bruce Rothnie (Deputy Surveyor of the New Forest) for granting permission to access Crown lands during lockdown; Esta Mion (Communications Manager); Andy Page (Head Keeper); and New Forest Keepers Matt Davis, Lee Knight, and Sandy Shaw. I would also like to thank Martin Noble (retired Head Keeper) and his wife Julia.

I often rely on the good faith and experience of those who look after the wider landscape I'm working in. It would not be possible to do my job without such support and I am of course especially grateful to Andy Page for being so generous with his knowledge and time. People like Andy and his colleagues at Queens House, Lyndhurst, are the reason that places like the New Forest are still able to provide a home for both people and wildlife. Balancing the needs of humans with those of complex, fragile ecosystems is no mean task – a theme that I hope this book goes some way towards highlighting.

Sincere thanks also to Matt Roseveare whose unfailing reliability and cheerful company took the edge off all those 3 a.m. starts and was crucial to the success of everything we attempted; and Manuel Hinge who knows the forest like few others and has been extremely generous with his time and knowledge this past year.

Thank you also to Lloyd and Rose Buck for sharing their knowledge, experience and passion for goshawks. Lloyd and Rose are superb at what they do – the very best – and you can see them and their birds in action on their website: www.lloydbuck.co.uk

I should perhaps also add that the contents of this book in no way represent the sentiments or official position of any persons/organisations mentioned (including Big Wave, Smithsonian, Terra Mater and the Forestry Commission etc). I am a middle-aged, middle-class male prone to all the usual idiosyncrasies, faults and grumbles typical of my kind and the diary format is an inherently subjective style of writing. So, if there are any anomalies or inaccuracies contained herein: *mea culpa* and the fault resides with me alone.

Special acknowledgements

My father passed away from pancreatic cancer in the early hours of 24 November 2020. Our nation was in the middle of its second lockdown, but thanks to the support, empathy and selfless dedication of the East Dorset Community Nursing Team, we were able to honour his final wishes and keep him at home with his family.

Those long, dark nights were tough, heartbreaking, and soul-destroying. Cancer is a cruel, treacherous and unforgiveable disease; but the way the NHS nursing team went above and beyond to provide help was both humbling and heroic. It was also surprisingly life-affirming: thank heavens for people like them and for organisations like the NHS and the various charities that provide critical care for those who need it most. Everyone deserves the right to see out their days with dignity, and the compassionate support my father received from the palliative care team showed me what it is to be truly human.

But it also showed me how underfunded and overstretched our nation's care support teams are. They need all the help we can give, which is why I'm donating a portion of the royalties from sales of this book to Marie Curie (Charitable Trust Number 207994) in my father's memory. I daresay it will be a drop in the ocean, but what else can we do in the face of an economy that places more emphasis on corporate financial gain than it does on public-sector healthcare.

Finally, above all, I would like to thank my family both here in the UK and in Australia. 2020 was a year of terrible extremes for millions around the world and I count myself blessed to be sharing my life with you.

ABOUT THE AUTHOR

James Aldred is an Emmy Award-winning documentary wildlife cameraman and filmmaker and the celebrated author of *The Man Who Climbs Trees*. He works with television and production companies around the world, including the BBC and National Geographic. He has collaborated with Sir David Attenborough on numerous projects including 'Life of Mammals', 'Planet Earth' and 'Green Planet' and been nominated for BAFTA/RTS awards many times. He was lucky to spend the national lockdown of spring and summer 2020 filming in the New Forest.